GREAT CEN[illegible]

EAST OF SHEF[illegible]ELD

Volume 1

Geoffrey Hurst

ISBN 0-947796-00-2

Published by Milepost Publications, 39, Kilton Glade, Worksop, Notts. S81 0PX.
Printed by Netherwood Dalton Ltd., Bradley Mills, Huddersfield.

Introduction

This book covers the former Great Central railway lines east from Sheffield Victoria through Worksop; Retford; Gainsborough; Brigg; to Barnetby, Clarborough Junction to Sykes Junction (Leverton branch), Lincoln, Market Rasen and Barnetby.

The Great Grimsby and Sheffield Junction railway (GG&SJ), Sheffield and Lincolnshire Junction (S&LJ) and Sheffield and Lincolnshire Extension (S&LE) railway companies presented schemes to parliament for most of the lines covered by this book. Before they were completed, they merged with Sheffield, Ashton under Lyme and Manchester Railway (SAM) to form the Manchester, Sheffield and Lincolnshire Railway (MSL). The line from Sheffield to Barnetby formed part of the mainline from Manchester to Grimsby. The MSL changed its name to Great Central Railway (GCR) upon completion of the London extension, the Grimsby line was demoted to a cross country line. Under the Railway Grouping Act, the GCR formed part of the London and North Eastern Railway (LNER) from 1923. Nationalisation in 1948, brought these lines under the control of the Eastern Region of British Railways.

Passenger traffic originally was not as great as expected and developed slowly. Freight traffic consisted of agricultural produce from Lincolnshire and coal, coke, iron and steel from Sheffield. The coal industry expanded slowly from Sheffield to just east of Worksop. The last colliery was sunk in 1929-30 and at present only 3 collieries out of 14 remain, none west of Kiveton Park. Through the 1950s, 60s into the 70s goods traffic declined from wagon loads to bulk loads. Only Worksop has retained its goods depot, with exports of refractory bricks, rail clips and timber to Shotton. Sand traffic returned to the Glass works after an absence of three years.

It was not until 1959 that any line closed, and today passenger trains operate between Sheffield, Worksop, Retford to Lincoln and Cleethorpes via Brigg and Lincoln to Cleethorpes via Market Rasen. The mainstay of freight traffic is coal from South Yorkshire, North Derbyshire and Nottinghamshire to Cottam and West Burton Power stations, while the Lincoln to Barnetby line carries oil and petroleum based products to and from London and Midlands to South Humberside. The part of the old mainline between Gainsborough and Barnetby is under threat of closure, a classic example of neglect with a poor service and little or no publicity.

Acknowledgements

I would like to acknowledge the following people and organisations who have made the production of this book possible. Nottinghamshire County Leisure Services at the following libraries :- Nottingham Central, Newark, Retford and Worksop, Lincolnshire Libraries at Lincoln and Gainsborough, Humberside Libraries at Grimsby and Scunthorpe, Doncaster Library, Sheffield City Libraries and Manchester City Libraries for the use of there photographs, maps and local newspapers, Nottinghamshire, Lincolnshire and West Yorkshire Archives, Public Record Office, National Railway Museum Library, British Coal Headquarters Library, British Rail (Eastern Region), British Steel Corporation and lastly the Central Electricity Generating Board. Photographers R.Allsop, W.A.Camwell, H.C.Casserley, G.Coltas, J.Edgington, S.Gay, M.A.King, L.A.Nixon, K.Pirt, D.Sutcliffe, Lens of Sutton, Heyday Publishing. Special thanks to H.Answer, Richard Allsop, Jim Jackson, Peter Fox, Glyn Waite and Jim Slater for the use of tickets, photographs, pamphlets and for the help they have provided. Lastly to the late Guy Hemingway historical notes which have helped me.

Geoff Hurst
October 1989

Bibliography

The following local newspapers Lincolnshire Chronicle, Stamford Mercury, Lindsey Star, Retford Times, Worksop Guardian, Nottingham Journal, Doncaster Gazette and the Sheffield Independent.

Bradshaws Timetables, and Great Central, London & North Eastern, British Rail Public and Working Timetables Working timetables.

Minute books of the Board of Directors, Manchester, Sheffield and Lincolnshire Railway; Board and Traffic Committee Minutes of Great Central Railway; Works and Traffic Committee Minutes of London & North Eastern Railway, and the various Wartime Railway Executive committees and lastly various British Rail Eastern Region papers.

(Title page) LNER Q4 (ex GCR8A) 0-8-0 5965 stands at an up signal between Shireoaks West Junction and Tranker Wood Crossing in 1927, with an eastbound mixed goods train, waiting to gain access to Worksop Yard. *(P.Ransome Wallis, National Railway Museum)*

(Front Cover) **GC Director 62662 'PRINCE OF WALES' eases the 2.16 p.m. local from Cleethorpes to Sheffield Victoria round the famous Whisker Hill curve past Sutcliffes Maltkiln in September 1959.** ***(P.J.Hughes, Colour–Rail BRE 373)***

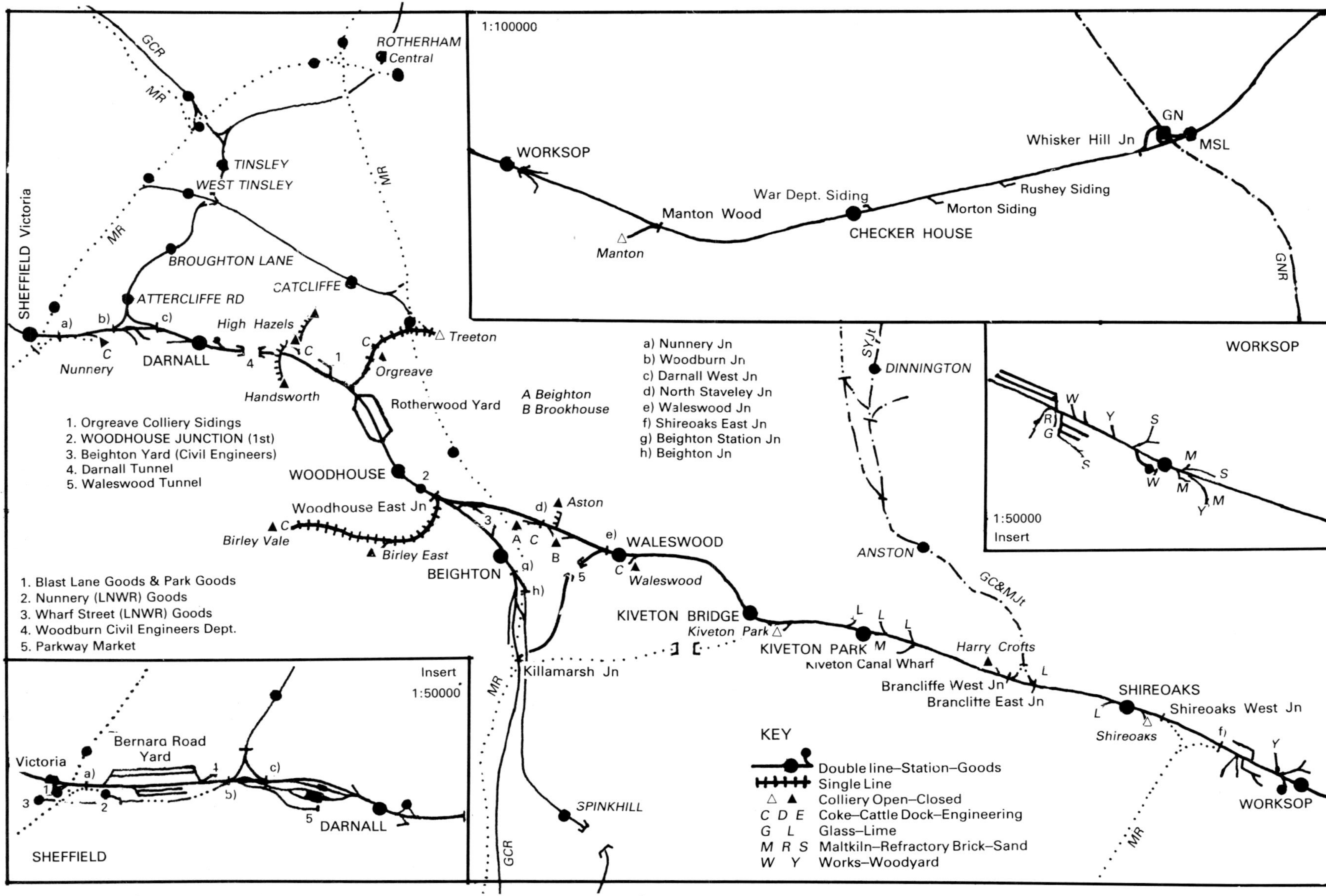
1:100000
WORKSOP
Manton
Manton Wood
War Dept. Siding
CHECKER HOUSE
Morton Siding
Rushey Siding
Whisker Hill Jn
GN
MSL
GNR
WORKSOP
1:50000
Insert
ROTHERHAM
Central
GCR
MR
TINSLEY
WEST TINSLEY
BROUGHTON LANE
ATTERCLIFFE RD
CATCLIFFE
SHEFFIELD Victoria
Nunnery
DARNALL
High Hazels
Handsworth
Treeton
Orgreave
Rotherwood Yard
A Beighton
B Brookhouse
a) Nunnery Jn
b) Woodburn Jn
c) Darnall West Jn
d) North Staveley Jn
e) Waleswood Jn
f) Shireoaks East Jn
g) Beighton Station Jn
h) Beighton Jn
DINNINGTON
SYJt
1. Orgreave Colliery Sidings
2. WOODHOUSE JUNCTION (1st)
3. Beighton Yard (Civil Engineers)
4. Darnall Tunnel
5. Waleswood Tunnel
WOODHOUSE
Woodhouse East Jn
Birley Vale
Birley East
Aston
WALESWOOD
Waleswood
BEIGHTON
ANSTON
GC&MJt
1. Blast Lane Goods & Park Goods
2. Nunnery (LNWR) Goods
3. Wharf Street (LNWR) Goods
4. Woodburn Civil Engineers Dept.
5. Parkway Market
KIVETON BRIDGE
Kiveton Park
KIVETON PARK
Kiveton Canal Wharf
Harry Crofts
Brancliffe West Jn
Brancliffe East Jn
SHIREOAKS
Shireoaks West Jn
Shireoaks
WORKSOP
Killamarsh Jn
SPINKHILL
Insert
1:50000
Victoria
Bernard Road Yard
DARNALL
SHEFFIELD
KEY
Double line–Station–Goods
Single Line
Colliery Open–Closed
C D E Coke–Cattle Dock–Engineering
G L Glass–Lime
M R S Maltkiln–Refractory Brick–Sand
W Y Works–Woodyard

1. Sheffield – Gainsborough

Plans were submitted in November 1844 for a railway from Sheffield, east to Worksop, Retford, Lincoln and Gainsborough. At Gainsborough (actually Bole on the west side of the River Trent) the railway would connect with The Great Grimsby and Sheffield Junction Railway. It appears that a field at Bole was chosen as a junction, as the Great Grimsby Act had a greater chance of being approved. Two branches were planned, both were to connect with the Midland Railway. The first branch was to Beighton, and the second was from Woodhouse Mill onto the mainline, by a north to east curve.

The Sheffield and Lincolnshire Junction Railway Act survived virtually intact to receive Royal Assent on 3rd August 1846. The act allowed the following :- Making a railway, commencing by junction of Sheffield, Ashton under Lyme and Manchester Railway (SAM) at or near Osbourne Street in the Parish of Sheffield and shall terminate in the said Parish of Beckingham at or near a certain bank or wharf adjoining the River Trent; and one of the said intended branch railways shall commence by a junction with the line of the first mentioned railway, in the Parish of Handsworth and shall terminate by a Junction with the line of the Midland Railway in the Parish of Beighton; and the other of the said intended branch railways shall commence by a Junction with afore mentioned railway, in the Parish of Aston cum Aughton, and shall terminate within the said parish by a Junction with the line of the said Midland Railway.

The company authorised to raise £700,000.

Level crossings in the parishes of Torksey 1, Saundby 1, Brightside Bierlow in the parish of Sheffield 1; Attercliffe cum Darnall 1; Aston cum Aughton 1; Wales 1; Shireoaks 1; Haggonfield 1; Worksop 1; Babworth 2; Ordsall 2; Clarborough 3; Sturton le Steeple 3; West Burton 1 and Beighton 1.

Junction with SAM, under direction of engineer of SAM.

A period of seven years allowed for construction.

Great Northern Railway may use portion of Sheffield and Lincolnshire Junction between Sheffield and if sanctioned by next 2 parliaments, at a rate exceeding 3 miles between Gainsborough and Retford.

Midland Railway to use of line between junction with proposed Manchester - Lincoln Union Railway near Worksop and terminus at Gainsborough.

Contract No.1 from the junction with Sheffield, Ashton under Lyme and Manchester railway to the junction with the Midland Railway at Beighton was let on 18th October to Miller & Blackie, with a tender of £66,000 and to be completed by April 1848. The tenders for contract numbers 3 and 4, from Woodhouse to the junction with the GG&SJ were to be in by 10th February 1847.

The '*Sheffield Independent*' said in the 27th March edition that :- "Last contract on S&LJ, for Wicker Viaduct, let on 23rd March to Miller & Blackie for completion April 1848. It is rather more than half a mile long and 50 foot wide, with space for 4 lines of rail. Will commence at the present Bridgehouses station, crosses just below Johnson Street, and crossing the Wicker, the River, Blonk Dam, Sheaf Works and the canal and will terminate on the hillside just beyond Soap House Pit. There will be two arches across the Wicker, one to accommodate the road traffic on Barnsley Road, the other the Midland Station and Saville Street. The work will commence with the erection of the scaffolding across the entire valley".

Reports appeared in the '*Sheffield Independent*' of 13th April, that - "On Wednesday, the rest of the Sheffield & Lincolnshire Junction (S&LJ) contracts except Sheffield Viaduct was let, Contract No.3 Woodhouse - Retford to Hutchins & Brown (£119,923), Contract No.4 Retford - Bole to Waring & Son (£112,000), for completion in October 1848. Sheffield Viaduct contract to be let within 6 weeks. Works let within £9,000 of parliamentary estimates. Chairs and rails let on same day to Scolefield & Co. of Parkgate, and sleepers let to Singleton & Harrison of Grimsby. Contracts let for a large number of wagons. During last month, surveys proceeded on S&LJ, gradients in Notts easy, tunnel of 550 yards at Welham Winnies 2 miles 70 chains northeast of Retford, with 51 feet cover at western end and 61 feet for eastern end, greatest cover 101 feet 6 inches for about 200 yards".

During the next month the '*Doncaster Gazette*' and '*Sheffield Independent*' reported the progress of the S&LJ. The 16th April edition said that the S&LJ works at Retford had commenced forthwith. The residence of Robert Millington, has been purchased as the site of Retford station. On week later, a report stated :- "A number of houses have been pulled at the eastend of Walker Street, Wicker adjoining the SAM station. To allow the start of the line across the Wicker and cross to Darnall". The next report appeared in the '*Doncaster Gazette*' of 25th June :-"The S&LJ is making considerable progress between Retford and the North Midland line, and near Retford. The tunnel at the Winnies, through Clarborough hill is advancing. Six shafts have been sunk, and tunneling has commenced".

The board were told on 18th August that :- Contract No.1 Sheffield Victoria rapid progress, Contract No.2 Excavation more than half complete, culverts completed, bridges in advance state. Only 200 yards remain at short tunnel at Handsworth. One and a half miles of permanent way laid at the west end of Handsworth Tunnel the contractors have quarried stone to be used on Sheffield Viaduct. Contract No.3 Rother Viaduct commenced and purchased land to Worksop, Worksop to Retford works light and land not yet obtained, a considerable portion of the line to Worksop excavated. Contract No.4 All shafts for Clarborough Tunnel excavated to full depth, and engineman and machines erected. The bottom headings are nearly complete, and excavation on both sides of the tunnel proceeding well. Bricks to be used are being made near Retford. Trent bridge at Gainsborough was subject to interference from the Admiralty, which caused considerable delay. However work is proceeding rapidly. Strike on S&LJ is over, all the tommy shops are to abolished, and the men are to draw 2s a day subsistence money, and settle every month.

An engineers report was reproduced by the '*Sheffield Independent*' on 19th February 1848, :- Contract No.1 Wicker the embankment east of the viaduct, 10,000 cubic yards finished, 8,000 cubic yards remain out of the 28,000 excavation for foundations. 20,000 rubble masonry brick, 50,000 remain. 70,000 cubic feet ashlar done, 190,000 remain. Contract 2. 490,000 cubic yards excavated, 110,000 remain. 8 bridges completed, remaining four in aforward state. 60 yards of Handsworth tunnel remain. 2½ of double track of permanent way laid, 4 miles remain. Contract 3. Rother Viaduct in good position and foundations are in, and the piers are nearly all half height. Arrangements have been made, for supply of bricks as soon as the weather permits for work to safely to proceed. 370,000 cubic yards excavation removed, 660,000 remain. Bridges from Rother Viaduct to Dog Kennel are in forward state, but only commenced thence to Worksop. The rest are small. Contract 4. 120,000 cubic yards excavated, chiefly from the west end of the tunnel, 430,000 cubic yards remain.Over 100 yards of Clarborough Tunnel finished, the bricks made at Retford renowned for their excellent quality, and the brickwork that is been done is very creditable to the contractor.

The construction of the Idle bridge at Retford caused problems and the '*Sheffield Independent*' of 22nd April, gave the following :- "At the Idle bridge, the ground is completely bog, and after the workman had dug a full several foot for the foundation, it was found so insecure, that it was deemed necessary to pile the work. Over 600,000 bricks had been used, the key brick was laid at 3.O'clock on the 14th by Mr Bullivent Jr. of Retford". Three days later they recorded that the "First stone of Manton Viaduct laid on Sat. 13th by Mr R.A.Pearce. of the Lion Hotel Worksop".

At the board meeting held on 22nd June, it is minuted that the tenders for rails and sleepers received. Plans for stations between Sheffield and Gainsborough approved, tenders for these and the cottages to be obtained.

The 12th August edition of the '*Sheffield Independent*' reported the MSL half yearly meeting :- "It has been arranged to open the branch to Midland at Beighton simultaneously with the Lincolnshire lines. The viaduct progresses well. It may be possible to open Sheffield to Beighton next month, Contract No.3 Rother viaduct well advanced and will be finished by the winter. Earthwork nearly completed thence to summit cutting at Wales only one fifth of the cutting still to do, thence to Tanker Wood near Worksop, earthworks finished and formed, little to do beyond. Contract No.4 Masonry nearly done. about 500 yards of Clarborough Tunnel remain to be finished. Somewhat extensive repairs have been required on Chesterfield Canal".

The Eastern Executive decided on 19th September 1848 they should build a siding to connect with Chesterfield Canal at Anston Bridge (just east of Kiveton Park). They authorised a wharf wall to be puddled, at a cost of £120. A shed to be covered with blue slate and a small crane at an estimated £235, if covered with wood the price would be reduced by £35.

Disaster struck on 30th September, when Rother viaduct collapsed. The '*Sheffield Independent*' of 7th October gave the following account :- "The viaduct crosses the Midland Railway above Woodhouse Mill by means of a viaduct of a quarter of a mile in length. About a mile from the point where the Lincolnshire crosses the Midland line, a branch from the main line (known as the Beighton Branch). The works on this portion of line form let to Messrs. Miller, Blackie and Shortridge, and by them underlet to Mr Carlisle. The viaduct, which is built upon a hard clayey foundation, consists of 36 arches, supported by strong and substantial built brick piers. They were all semi–arches, each of 30 feet span, and an average height of 20 feet from the surface to the spring: the height from spring to the crown averaged 15 feet. The work had, under Mr Carlisle's superintendence, proceeded satisfactorily up to the completion of the twentieth arch, and from a portion of those which were first finished, the centres had been withdrawn. The works were visited daily and the substantial character of the fabric had elicited there marked approval. The late heavy rains, however, in some measure, interfered with the progress of the works; and a fortnight ago it was observed that the piers of the nineteenth arch, from which the centres had been eased, had sunk more than usual in cases where the foundations were secure. Levelling pegs were put down, and it was ascertained that there had been settling of the whole of the piers since their commencement, nowhere, except the nineteenth arch, which was found to have sunk 14 inches, owing, it is supposed to some soft under strata.On Friday, it was deemed necessary to stay the arches near the nineteenth, in order, if possible to prevent the falling of the whole, if that (the nineteenth) resting upon defective foundations. During the whole of Friday, seven or eight men were employed upon this work. When the first great sinking of these piers was observed, steps were taken by the engineers to ascertain the cause of it, but the heavy falls of rain prevented their proceeding in the investigation. On Friday, the River Rother overflowed its banks for some distance, and in some places the meadows were flooded to a depth of four or five feet. Operations of the workmen resumed on Saturday morning, and a few minutes before ten o'clock, the nineteenth arch, under which four men named Henry Whiteman, Geo. Belcliffe, James Mott, and William Lamb were at work, fell in upon them with a loud crash, killing two of them (Whiteman and Belcliffe) on the spot, and injuring Mott so severely that he died a few hours later. Lamb survived with a fractured arm and other serious injuries. The first arch fell with a crash like a report of firearms. The others fell in one by one in quick but distinct succession, causing alarm in the neighbourhood. Later in the day two or three arches which were keyed and all but complete, also fell in. The mass of rubble caused a dam effect on the River Rother increasing the flooding to rise upto fifteen feet above the river bed".

Soon after the collapse of the Rother Viaduct, the Wicker/Sheffield Viaduct partially fell down. The '*Sheffield Independent*' of 17th October gave the following report :- "Fall of the arches of Wicker Viaduct, not to a serious extent as Rother Viaduct, and unattended by the lost of life. 700 yards in length extending from the present station in Clay gardens having its abutments on the east of the canal. That portion of the viaduct Wicker and Effingham Lane is greatly increased width to afford room for a passenger station about 300 yards in length. From Effingham Lane to termination on the eastside of the canal there are six flat segmental arches, two of them being 50 feet , and the others 47 feet spanned the later having a rise of 10 foot. From the eastend, first arch crosses the road by the canalside, second canal, and the third (the arch which has fallen) crosses a narrow private road between yard of the Sheaf Works and the canal, and enters upto 20 foot into the Sheaf works yard. The arches are of brick faced with each side by a row of stone quoins. The arch on each side commences with seven rings of ridges of bricks reduced six, and to five at the centre, the centres have had been removed, after several weeks, and the asphaltum covering had been laid. At about 11 o'clock on Wednesday morning, the 4th third arch suddenly fell, and several persons had narrow escapes. Restoration work is proceeding day and night".

The Eastern Executive Committee inspected the works under its control over a period of three days from 9th November. They reported the following findings :- "Darnall work let and wood work prepared, but nothing further done. Same at Beighton, Dog Kennels (Kiveton Park) and Shireoaks. Kiveton Wharf let to Messrs. Wright. Worksop Goods Shed plans show 220 feet by 60 feet, cost £4,890. Present wants not require length, recommend that half be boxed up! Coal shoots finished except slides. Retford station to be curtailed, by doing away with refreshment room and goods warehouse, at present only half built. The coal shoots to be suspended. Gainsborough station commodious".

The '*Doncaster Gazette*' of 10th December 1848 described progress :- "The heavy works in the Retford area nearing completion, nearly eight miles being ready for ballast and permanent way. Clarborough tunnel, 550 yards is it open throughout except at the entrances.This passes through a very hard marl, interspersed with veins of gypsum, lying on a flat bed, varying from 1 to 6 inches in thickness, brilliantly white and sparkling. Not so much water that except meet with as had been expected. Dominis Cross cutting, 27 foot deep, is nearly finished. Across the Idle valley there is width for five track, to accommodate the GN. The foundations of Retford passenger station have been put in and the walls are upto rail level. The foundations of the goods station are being laid".

The following month the committee minuted that :- "The junction with the Midland Railway at Beighton would be simple. The only condition was that the Manchester, Sheffield and Lincolnshire Railway must pay for a watching station to be controlled by the Midland Railway".

The '*Sheffield Independent*' newspaper announced "We are glad to state that the line from Sheffield to Beighton, where the junction takes place with the Midland, is so far completed that a trial trip was made upon Saturday last. The Hesperus engine, driven by Mr W.Holmes, manager of the locomotive department, started from Bridgehouses station at half past two, and ran to the junction in 19 minutes, including stoppages. It is expected that this line will open during January, and by arrangement with the Midland railway these trains will run to Eckington. Village stations are being erected at Darnall and Beighton".

This proved to be to hopeful, as the same newspaper reported on 10th February 1849 :- "The line from Sheffield to Beighton was inspected yesterday by Capt. Wynne. The portion of the line is in length 7¼ miles, and the works upon it are heaviest in the eastern section of the railway. It commences at the Bridgehouses station by an immense viaduct, 660 yards long, crossing the Wicker, River Dun, and the canal, and terminating on the hill on the eastern side of the canal. The height of the viaduct over the river is 57 feet. The line then runs over embankmentsand through cuttings alternately, ascending generally at a rate of 1:150 to Handsworth Tunnel. This tunnel is 374 yards long. From it descends by 1 :150 to the junction with the Midland. The highest embankment is 51 feet. The stations on this part of the line are at Darnall and at Beighton. The viaduct cost £80,000. More than half of this outlay has been incurred, not for the viaduct solely, but with a view to the proposed passenger station on the site of Blonk Dam, which, though deferred for the present, must be ultimately adopted. The cost of the works on the other part of the line has been £13,000 per mile.

Capt. Wynne made his inspection accompanied by several directors and officers of the company. When the Vesta engine, with two carriages, had passed over the viaduct, Capt. Wynne and the company got into the train, and proceeeded slowly, Capt. Wynne stopping to inspect every bridge, and walking through the Handsworth tunnel. At Darnall, a small wooden station has been constructed. The Orgreave road crosses the railway by a timber bridge, because of mine workings below. The Worksop turnpike road is crossed by a bridge of wrought iron tubular girders. This was required to meet parliamentary requirements. Approaching Beighton, the branch to the Midland leaves the mainline on a embankment. At Beighton a neat brick station is nearly completed and well finished. After a short stay at Eckington the party returned to Sheffield in 14 minutes."

Table 1

Weekdays			TSO						Sunday			
	am	am	am	am	pm	pm	pm	pm	am	am	pm	pm
MANCHESTER											*9 0*	
SHEFFIELD	645	815	945	1140	1 0	145	415	545	730	840	430	545
DARNALL	652	822	952		1 7		422	552	737	847	437	552
BEIGHTON	7 3	833	10 3		118		433	6 3	748	858	448	6 3
ECKINGTON MR	713	845	1013	12 4	128	2 9	443	613	758	9 8	458	613

Weekdays		TSO							Sunday			
	am	am	am	pm	pm	pm	pm	pm	am	am	pm	pm
ECKINGTON MR	728	9 0	1047	1210	148	3 2	5 5	655	813	930	5 5	655
BEIGHTON	740	912	1059	1222				7 7	825	942	517	642
DARNALL	751	923	1110	1233				718	836	953	528	653
SHEFFIELD	756	928	1115	1238	211	325	528	723	841	958	533	658
MANCHESTER	*1024*		*139*						*1139*		*839*	

From Manchester

	Fares		
	1st	2nd	3rd
Sheffield	7/6	5/6	3/5
Darnall	7/9	5/9	3/7½
Beighton	8/6	6/3	4/0
Eckington	9/0	6/8	4/4

From Eckington

	Fares		
	1st	2nd	3rd
Beighton	1/0	0/8	0/4
Darnall	1/9	1/3	0/8½
Sheffield	2/2	1/7	0/11
Manchester	9/0	6/8	4/4

The MSL half yearly meeting of 13th March 1849 were told :- "Contractor provided centres for every arch of the Rother Viaduct, so work can proceed rapidly. All the piers are at springing level. Completion expected early May. The stations are proceeding well, Sheffield station has been suspended for sometime, there is inconvenience in working the Beighton traffic to the old station at Bridgehouses. Clarboro Tunnel is complete and the small amount of excavation at either end, 50,000 cubic yards remain in Manton cutting, and about the same amount at Tranker Wood. About 20,000 cubic yards remain in the summit cutting, at Wales".

On 20th April the *'Doncaster Gazette'* reported that the first engine on the (Gainsborough) line had pulled 12 wagons of bricks to Manton.

During 1849 the Eastern Committee decided that they required a new station at Sheffield to allow through running from Manchester to the eastern lines.The committee were informed on 25th April that an estimate of £25,000. The MSL Board were told on 2nd May by letter from Mr Fowler that Messrs Waring (the contractors) absurd and unjustifiable conduct, is likely to delay the opening of the line on 1st July. Mr Fowler looked for a new contractor to take over the Waring Contract. Messrs Waring applied for an injunction to prevent the MSL taking Contract No.4 into their owns hands. On 29th June the MSL board decided that a committee be formed to deal with Great Northern matters. They were told that an agreement had been made for the MSL to conduct GN traffic at Sheffield and the GN to conduct MSL traffic at Lincoln on 13th June.

The Eastern Committee met on 18th July 1849, they wrote a letter to Mr Fowler asking - "Why Worksop station roof put up without orders from the committee, after they had been suspended execution of work sometime ago, and had any further orders been given in regard to Retford Station". At the next meeting Mr Fowler replied and they noted the letter, but did not disclose its contents, which appeared to satisfy them. A memorandum was sent to the committee, from the inhabitants of Handsworth praying for a station at that place in lieu of Darnall. The committee on 1st August asked for estimates to be prepared.

The line between Woodhouse and Gainsborough opened on 18th July 1849 completing the final section of line between Sheffield and Grimsby to form the east to west mainline.

The *'Sheffield Independent'* correspondent describes the line as follows :- "The Sheffield Viaduct is the greatest work of the Lincolnshire Line. It commences immediately on leaving the Bridgehouses station for the eastward and stretches a length of 660 yards across the valley of the Dun. Its greatest stride is by one magnificent arch of 72 feet span across the Wicker road. It then crosses the river and the canal, till it reaches the high ground on the east bank of the canal. This viaduct contains a greater amount of masonry by far than any similar work in England. Of its magnitude, the following figures give a faint idea. The foundations amount to 34,000 cubic yards. There are 70,000 cubic yards rubble work, and 250,000 cubic feet of ashlar masonry. It is calculated that there is in this viaduct as much masonry as in a thousand turnpike road bridges, or seven hundred large parish churches. In addition to the viaduct, piers have been built for a station, covering about four acres. This is now in progress, and is expected to be so far completed as to be open for passengers in about six months. The Sheffield viaduct has been built by Messrs. Miller, Blackie, and Shortridge, and was commenced in 1847. Notwithstanding the delay occasioned by financial pressure of the times, it was ready in about nineteen months. The viaduct had cost up to the time of its opening about £80,000, but more than half the outlay had been incurred, not for the sake of the viaduct proper, but with a view to a passenger station. The facility for obtaining stone has rendered this great work remarkable for its cheapness.

Leaving this viaduct, the only work of importance before we reach the Rother viaduct, is the Handsworth tunnel, 374 yards in length. To the first cutting on leaving Sheffield, the line passes over broken ground of very uninviting appearance. But looking to the north side of the valley, there is a fine view rich country from Hallcar to Meadow Head, embracing the bold wooded hill of Wincobank, where may traced the remains of a Roman camp and a Roman ridge extending northwards to Hall Hill. On emerging form the first cutting, about two miles from Sheffield, we have a short run through a very beautiful country, leaving on the southside the village of Darnall and Tinsley Park. We lose all of this at High Hazels, by plunging into the tunnel, where we emerge to enter the parish of Rotherham, through Highfield Spring wood.This brings us into the fine valley of the Rother, remarkable for the richness of its scenery and the fertility of its soil. Here are many points of interest in the scenery. The pretty village of Beighton, the Rother Viaduct, which curving towards the Midland line, may be seen to great advantage from the carriage windows, the villages of Woodhouse, Hackenthorpe, etc. The Rother Viaduct was made the subject of a full examination by Capt. Wynne. The disastrous fall of 21 of its arches in September 1848, assisted by the premature removal of one of the centres, naturally made him very careful to ascertain that the best means had been adopted to render impossible the recurrence of such a catastrophe. Since the fall, the piers have been rebuilt upon piles, and the whole work has been so solidly constructed as to give Capt. Wynne the most perfect satisfaction. The viaduct consists of 36 arches, of 30 feet span, the average height 40 feet. It forms a noble piece of work, and is remarkable for the rapid and yet substantial manner in which it has been completed. The viaduct cost £17,000.

Leaving the viaduct, we pass through country of great beauty till the line enters the great Wales cutting. This the ridge through which the Chesterfield canal passes by a tunnel, the railway, by a cutting of 400,000 cubic yards. This part of the line is one of special interest to the geologist. West of the cutting the line runs through strata intersected by coal measures. In the cutting we find a thin stratum of the lower new red sandstone, and again we pass from it to the magnesian limestone. This continues to be the character of the stratum, as far as Brantcliffe, on the borders of Nottinghamshire, when we come again upon the new red sandstone.

The railway leaves the cutting at Kiveton Park Station, hereto afore better known as the Dog Kennels. Here it meets the Chesterfield Canal, which at the same point emerges from its tunnel. The celebrated Anston stone which has been brought into so much repute by its adoption, as the stone for the building of the New Houses of Parliament, is got about a mile of this place, and here the boats were loaded to convey down the Trent to the banks of the Thames. On one side the canal is Mr Grizzell's Wharf, for shipping the parliamentary stone; on the other, that of Messrs W. Wright and Co., who are working the Anston quarries. This bids fair to be a busy station. The stone quarries demand large means of transit, and there are here also lime kilns doing considerable business. Other limekilns are about to be built. The neighbourhood also is agreeable and attractive for residence, and we are informed that applications have already been made to the Duke of Leeds's agents for sites to erect villa residences.

The next station is at Shireoaks, and here the attention of Capt. Wynne was directed to the improved semaphore signals on the portion of the line about to be opened. The improved construction has had two advantages. It causes one lamp, having a face each way, to supply the place of two; and, secondly, it adds to the signal given the light, red, green, or white, that of a wooden arm, worked by the same lever as the lights, so that the arm and the lights can never through carelessness in the signal man contradict each other. When the red light is shown, indicating danger, the arm is stretched out, as it to bar passage. When the green light shows says to go cautiously, the arm stands obliquely from the post. But when the white shows that there is no obstruction in the way, the arm disappears from sight. By making one lamp do the work of two, having a face both to the up and down line, £15 a year is saved, and the first cost is also considerably diminished. Capt. Wynne expressed himself pleaded with the invention, and requested to have tracings sent to him, showing the working details of such an important and economical improvement.Mr Fowler explained to Capt. Wynne, that he had been indebted to our townsperson, Mr W.W.Cutts, for this improvement, and for superior workmanship by which it had been so successfully carried out; and he believed from the great advantages obtained, that 'CUTTS SIGNALS' would be universally adopted.

There are several portions of the line between Kiveton Park and Worksop where, for two, three, four and five miles, it runs in a perfectly straight line. The road near Worksop is crossed by a wrought iron tubular skew bridge, giving 65 feet 9 inches of clear opening, manufactured by Messrs. Fairburn, of Manchester. At this place, the heat of the sun directed attention to the newly laid asphaltum, which is used on all bridges, and viaducts on this line, to protect the works from water, and has been applied by Messrs. Wright & Co. It has two great advantages over the old mode of pudding with clay. It answers the purpose more effectually, and it does not encumber the works as clay does by adding weight without any increase of strength. Where ever the iron bridges occurred, Capt. Wynne was careful to test them, and ascertain the amount of deflection. It was found in every case to very insignificant.

Worksop station is architecturally, the most elegant upon the line. The design combines neatness, with all that is desirable in the way of ornament. It is built of Steetly stone, the quarries of which are about two miles away from Worksop, and are worked by Mr Carlisle. The characteristics of the stone are its beautiful whiteness, the ease with which it is worked. The Worksop station, seem in sunshine from a short distance, looks like white marble. Worksop will, no doubt, be a favourite destination for excursion trains and private pleasure parties. Within a short distance are Clumber, Welbeck Abbey, Thoresby, the charming glades of Burkland and Sherwood Forest, the dismantled Worksop Manor house, and, at Worksop itself, the ruins of the old abbey, the church of which has just been restored. A fine run of eight or nine miles across the highly cultivated enclosures and park like plantations of the forest brings traveller to Retford, where the line crosses the mainline of the Great Northern, uniting it by a short branch, which runs from the Great Northern onto the Manchester, Sheffield and Lincolnshire Company's station for the town of Retford. From Kiveton Park to Retford the works are generally light, and the appearance of the brick work is greatly set off by the appropriate use of Steetly stone.

The Retford station is destined to become of great importance. Here the string of the bow, formed by the loop of the Great Northern, is to intersect the Lincolnshire line on its way from Bawtry to Peterborough, giving to this part of Yorkshire the most direct and speedy route to London. This station has been built during the last three months, under the active management of Messrs. Hutchings, Brown, and Wright. It is built of white and red brick, with mountings of Anston stone. The last contract from from about three miles west of Retford to two miles east of Gainsborough, was let to Messrs. Warings and Sons, but, inconsequence of there not undertaking satisfactory progress, it was taken out of there hands, after a chancery suit, a short time ago, and has been completed with great effort by Messrs. Peto, Betts, and Giles. The Retford station is close upon the southside of the town, but about half a mile distant from the market place. Mr Dennett has provided omnibuses to run to and from the station. One of these is a new carriage, and, taken altogether, is the most tasty turnout of the kind we have seen for a long time. This leads us to mention the very superior carriages which the Railway Company work upon this line. The first class carriages especially are lofty, well ventilated, elegant, and comfortable far beyond anything we have previously had in the neighbourhood.At Retford the train stopped about an hour and a half, and Mr Fowler invited Capt. Wynne and the gentlemen in the train to dine with him at the White Hart. The party returned to the train at about four o'clock.

From Retford the line runs for most part upon the level, and with great straightness. There is, however a considerable embankment at Welham, and the Grove Hills are tunnelled through at Clarborough. These hills consist of a mound of gypsum, overlaying the new red sandstone, and the gypsum formation is well developed in the cuttings at each end of the tunnel, which is 656 yards in length.

The Retford and Gainsborough contract terminates at Bole, and here we enter on the Great Grimsby. Here we have the greatest engineering feature of the eastern section of the line. The Trent bridge is approached by on the western side, over a wooden viaduct, the only wooden viaduct in the Eastern section. Wood has been preferred in this case in order that the viaduct may afford the least possible obstruction to the passing off of the floods in this valley. The bridge is thus described - It is a hollow girder bridge,of the largest dimensions yet executed of similar constructions, and differs from the Britannia and Conway bridges in this respect, - that instead of the trains running through the inside of the tube, the girders form the parapets of the bridge, and the roadway is supported by transverse wrought iron hollow beams, also of tubular construction. This bridge has been designed by Mr Fowler, the engineer of the line, and the girders are upon Mr Fairburn's patent principle of construction, and executed by Messrs. Fairburn and Sons, of Manchester. The stone work consists of a centre pier, and a handsome elliptical arch of 50 feet span, on eachside of the river, terminating by substantial and appropriate land abutments. The masonry has been executed by Messrs. Stephenson & Co., the constructors of this part of the line, and reflects great credit upon those gentlemen. The bridge is composed of two spans of 104 feet each, width, with the two land arches and abutments, give a total length of about 400 feet. The dimensions of the principal girders are 336 feet long, 12 feet high, and 3 feet 1 inch wide. The tops of the girders are formed of two cells; 18 inches wide and 12 inches deep, to resist compression, and the bottom of the double rivetted plates, to resist tension. The girders are fixed securely on the middle pier; on the land abutments there ends are supported upon rollers, resting on cast iron pates bedded into the masonry, to admit of there contraction and expansion. On the outside of the girders are rivetted two parallel lines of angle iron, in the form of an arch, which spring from the middle pier to the land abutments on each side, and gives a chaste and symmetrical appearance to the structure. The weight of the iron work in the bridge is as under :- The two principal girders, each 150 tons; transverse beams, four feet asunder 82 tons and Cast Iron 10 tons. Giving a total weight of 392 tons.

It was originally intended to have had them constructed on a platform or scaffolding, supported upon piles driven across the bed of the river; but the depth of the water, the nature of the ground, and the necessity of preserving the navigation open, rendered such a process inadmissible. Subsequently it was determined to construct them at one end of the embankment, and to haul them across upon rollers from one side of the river to the other; but much greater difficulties have been encountered than were originally contemplated, as the immense weight of the girders, when resting upon comparatively few points of support, caused the embankment to settle,and much time and trouble were expended before it was sufficiently consolidated to bear the load. The process of hauling the girders without any intermediate support, except a roller fixed upon piles at a few from the middle pier and last abutment, was a formidable undertaking.In this operation one of the girder had to be supported 130 feet before it received this auxiliary support. These operations were, however, successfully accomplished, under the direction of Mr Fowler and Mr Fairburn.

The testing of this great work was the most interesting part of Wednesday's proceedings. There were collected six engines with there tenders loaded with coke and water, besides several carriages, forming two ponderous trains of 200 tons, running upon the two lines of rails. These ponderous masses stood upon the bridge in various positions, while the amount of deflexion caused by the weight was tested,and it was found to be 1¼ inch. It was calculated that a weight of 500 tons would be required to depress it to a level. Capt. Wynne next took his station below to watch the effect of the trains passing again and again over the bridge. The vibration was only just perceptible. He next tested it by standing upon the piers of the bridge while the train passed over, and at every point, the result was most satisfactory. So far from there being the slightest weakness, there was, on the contrary, an evident super abundance of strength. Having spent a considerable period in testing this important work, the train ran forward to Gainsborough, which has a large and a very complete station, conveniently situated".

Having obtained Captain Wynne's consent to the opening, large posters were immediately issued announcing that the line would be permanently opened throughout for traffic on Tuesday.

Table 2

Weekdays						Sunday			Fares from Manchester		
	am	am	am	pm	pm	am	am	pm	1st	2nd	3rd
MANCHESTER		*6 0*	*1030*	*2 0*	*345*						
SHEFFIELD	615	8 5	1214	4 0	536	810	1025	4 0	*7/6*	*5/6*	*3/5*
DARNAL	621	811		411		816	1031	411	*7/9*	*5/9*	*3/7½*
KIVETON PARK	640	830		430		835	1050	430	*9/1*	*6/10*	*4/4*
SHIREOAKS	648	838		438		843	1058	438	*9/8*	*7/3*	*4/7*
WORKSOP	654	844	1248	454	610	849	11 4	454	*10/0*	*7/6*	*4/10*
RETFORD	712	9 2	1 6	5 2	628	9 7	1122	5 2	*11/5*	*8/7*	*5/5*
STURTON	726	916		516		921	1136	516	*12/4*	*9/4*	*5/10½*
GAINSBOROUGH	736	926	127	526	649	933	1144	526	*13/1*	*9/10*	*6/2½*
BLYTON	748	938		538		943	1158	538	*13/10*	*10/5*	*6/7*
NORTHORPE	756	946		546		951	12 6	546	*14/4*	*10/9*	*6/10*
KIRTON LINDSEY	8 3	953	150	553	712	958	1213	553	*14/10*	*11/2*	*7/0½*
SCAWBY & HIBALDSTOW	812	10 2		6 2		1007	1222	6 2	*15/5*	*11/7*	*7/4*
BRIGG	820	1010	2 4	610	726	1015	1230	610	*15/11*	*12/0*	*7/7*
BARNETBY JN	827	1017	211	617	733	1022	1237	617	*26/7*	*12/5*	*7/10½*
GREAT GRIMSBY *	*927*	*1133*	*248*	*711*	*813*		*125*	*711*	*19/3*	*14/6*	*9/2*
NEW HOLLAND	9 0	1050	240	650	8 2	1055.	110	650	*18/0*	*14/2*	*8/11*
HULL	935	1120	3 9	720	832	1125	140	720	*18/6*	*14/6*	*9/3*

* Change Ulceby

Weekdays						Sunday			Fares		
	am	am	pm	pm	pm	am	pm	pm	1st	2nd	3rd
HULL	*6 0*	*910*	*1245*	*315*	*545*	*815*	*315*	*545*			
NEW HOLLAND	630	940	115	345	630	850	345	630			
GREAT GRIMSBY *	*615*	*919*	*1248*	*315*	*540*	*838*		*540*			
BARNETBY JN	7 3	10 9	148	414	648	923	418	648	*2/7½*	*1/11*	*1/4*
BRIGG ↰	712	1018	157	423	657	932	427	657	*3/3½*	*1/11*	*1/8*
SCAWBY & HIBALDSTOW	720		2 5		7 5	940	435	7 5	*3/9*	*2/5½*	*1/10½*
KIRTON LINDSEY	729	1033	214	438	714	949	444	714	*4/4*	*2/9½*	*2/2*
NORTHORPE	736		221		721	959	451	721	*4/9½*	*3/2½*	*2/4½*
BLYTON	744		229		729	10 4	459	729	*5/3½*	*3/7*	*2/7½*
GAINSBOROUGH	756	1055	241	450	741	1016	511	741	*6/2*	*3/11½*	*3/0½*
STURTON	8 6		251		751	1026	521	751	*6/10*	*4/7½*	*3/4*
RETFORD	819	1117	3 4	522	8 4	1039	534	8 4	*7/10*	*5/1*	*3/10*
WORKSOP	838	1135	322	540	822	1057	552	822	*9/3*	*5/10*	*4/6*
SHIREOAKS	843		328		828	11 3	558	828	*9/7*	*6/11*	*4/8*
KIVETON PARK	851		336		836	1111	6 6	836	*10/2*	*7/2*	*4/11*
DARNAL	910		355		855	1130	625	855	*11/7*	*7/7*	*5/7*
SHEFFIELD	914	12 6	359	611	859	1134	629	859	*12/0*	*8/7*	*5/9½*
MANCHESTER London Rd	*11 7*	*156*	*6 4*	*756*		*139*	*834*		*18/6*	*9/0*	*9/3*

* Change Ulceby

The Eastern committee meeting of 30th July 1849, two weeks after the Woodhouse to Gainsborough section opened. Land was to be purchased from the Duke of Norfolk for construction of Sheffield Victoria station. Darnall station the committee do not apprehend they will be any ground to do away with it. Nothing at Beighton. At Woodhouse Junction a small station is required, but not to exceed £1,000. The crane pit at Kiveton Park was completed, but the platform was completed to half its length. A wooden shed at Thorpe Wood and crane indispensable is a stage for unloading of grain and goods onto canal to Chesterfield, crossing and siding to be put down. Shireoaks, distant signal required, but no further outlay necessary. Worksop, Retford and Gainsborough see estimates. At Sturton a small wooden shed and platform for commerce and cattle, coal, and lime, plus a short siding required.

On 25th October 1849, the committee decided that tenders for a cottage at Beighton should be obtained. Plans for Sheffield station were received. The engineer was instructed to prepare fresh ones, with the booking office to be reduced from 60 feet to 50 feet, also the offices to be reduced. Half shed up for length of building, 90 feet at end for in addition as a cover for cabs. During the last two months of 1849, the Eastern Executive committee decided that on 14th November, a siding for Bole should be referred to Mr Fowler. Three weeks later the subject of a station at Orgreave for Handsworth should be postponed. At the next meeting a week later they still regret Woodhouse station unfinished, to the extent, of necessary accommodation. The provision of facilities at Bole was postponed, until they knew the goods requirements of the district.Sturton station able to cop with the districts requirements.

Great Northern minutes say that they were experiencing problems with the MSL, who refused to supply water at Retford. The GN secretary wrote to the MSL on 12th March 1850, requesting that the order to stop supplying water be withdrawn. On week later the GN had received no reply. The GN board then decided to obtain an injunction to compel the MSL to do so. As a result of this action the GN General Manager suggested to the board that they build a station of there own. The MSL board minuted on 16th April 1850 :- Chancery Court case proceeding between GN and MSL. On 20th July, the MSL minutes recorded that they had lost the court action, and if the company did not provide facilities agreed upon then the company would be subject to sequestration, they instructed Mr Fowler to arrange with Mr Alport accordingly.

The *'Sheffield Independent'* informed its readers on 1st September :- "Sheffield station works have resumed, rapid progress is in the making preparation with that part intended for passengers. The small old station in Bridgehouses has been adapted as far as it would be accomplished at a moderate expense to through passenger traffic. The Sheffield Goods and Mineral stations should now be proceeded with. The MSL reduced the Lincolnshire line services from 5 to 4 on weekdays and 3 to 2 on Sunday, from April 1850. Yet the increase in traffic resulted as the 'Sheffield Independent' of 3rd May reporting :- "The MSL forced by increasing traffic, the necessity to improve arrangements in Sheffield. Arrangements made to move passenger traffic from Bridgehouses to new station on Blonk Dam, where the permanent platform and roof has been so provided for the emergency. Temporary booking offices will be erected, the way to the station will up the steps in the Wicker. Improved access will be made through the Cattle Market. The Bridgehouses station will be adapted exclusively for the use of goods and minerals".

The board gave orders on 28th December, for the prosecution of the station on Blonk Dam, with accordance of the design sanctioned by parliament. Roof of station of glass and iron after the fashion of Mr Paxtons glass palace. Expect roof in by April and open in May.

Sheffield (Bridgehouses) station closed to passengers on 14th September 1851., The new station at Blonk Dam opened the next day without any ceremony. The new station, allowed the MSL to run through trains to its eastern lines without reversing. The new station was later named Sheffield Victoria.

The MSLR submitted plans in November 1851 for various coal branches of which three were planned to leave the Sheffield to Grimsby line. The Manor Colliery branch of 65 chains was to have left the line from west of Darnall station, the junction facing east. The line descended 1:50 to the foot of the colliery. The second branch to Tinsley Park Colliery was to be 1 miles 23 chains long, leaving the main line by an eastern facing junction. The third branch to Birley, climbing 1:73 along the valley of Shire Brook for 2 miles 54 chains. All three lines were authorised the 1852 Act.

The Manchester, Sheffield and Lincolnshire Coal Branches Act of 30th June 1852, authorised the construction of the following :-

Railway No.2 Junction at Darnall Station, in parish of Attercliffe, terminate in colliery known as Manor Colliery.

Railway No.3 Certain wood called Highfield Spring, in township of Catcliffe, Parish of Rotherham. Terminate at or near tramway belonging to Tinsley Park Colliery, township of Tinsley.

Railway No.4 Beighton Branch, commencing parish of Beighton and Handsworth, terminating at or near a certain plantation reputed or belonging to Earl of Manvers, in the occupation of the said Earl.

Only Railway No.4 (Birley branch) was constructed. The directors approved the tender from T.Waring & Co. of £9,988 14s 5d on 28th April 1854. A report made on 2nd August, said work was progressing favourably and should be complete before the end of the year. This proved to be wrong as in the early part of 1855, the contractors stated that there work had been held up by problems in obtaining land. A board meeting held on 23rd March, were told the following :- Birley branch cuttings complete to new pits, except a small part in No.1 cutting, for ballast, permanent way may be laid by 16th April. Even before the branch opened, Birley Coal Company asked the MSL on 6th July for a siding, for its Coke Works, where Birley small coal was to be converted into coke. The board agreed to this request, estimating it would cost £60. The branch opened on 1st August 1855, and the maintenance contract for 12 months was awarded on 7th December for £320.

Mr Watkins recommended to the company board meeting held on 3rd April 1857, that it provide a weighing machine for lime at Kiveton Park, estimated at £160. To save shunting at Retford, the company were told that it would be necessary to alter the sidings, and was estimated to cost £1,600, also the water column caused problems for the operating staff, as it was necessary to uncouple the engine and back through the pits. The engine turntable needed moving to between cattle siding and standing siding. This is where it could be easily got at, either from the mainline, platform line or goods line. The turntable is unaccessible when shunting occurs.

The MSL asked for tenders for the construction of the Woodhouse north to east curve authorised in the 1846 Act. Three tenders were received on 8th July 1857 from Pocklington & Townend £5,182; Cameron & Nicholls £5,300 and J.Taylor £5,645 13s, the board decided to postpone the project and it was never proposed again.

A memorandum was sent to the MSL board by the people of Killamarsh, requesting on 20th January 1858, that the company provide a station. The board replied, they would stop at Killamarsh, if the Midland Railway would make a station. Mr Hargreaves made an application to the company on 21st May, for a wire communication between Worksop Goods and pointsmen box at level crossing. To enable communication with box, to save time, rather than working goods onto mainline. The wire was estimated to cost £18. Mr Bassett was to open a works for confectionery at Darnall, and he asked the company, to turn Darnall station into a goods station. The board received this request on 25th June, and they minuted that the Directors not in a position to expend any capital at present, but would be willing to give any other facilities.

The opening of the GN station at Retford presented operating problems for the London to Sheffield services which called at Retford, as they had to reverse out the station and proceed round the south curve. In order to alleviate this difficulty the GN suggested to the MSL that they build a curve from the Worksop line to the north of the GN station. The carrot for the MSL was a rent free station and the closure of the MSL station would save the company money. The board were told on 25th March 1859, that Messrs Newall would sell the land to the MSL for £290 an acre, but objected to the crossing of the turnpike on the level. The curve was quickly built and was inspected and sanctioned for passenger use from 1st July. From that day the MSL station closed to passenger, and all the companies trains stopping at Retford used the GN station. Retford south curve was now surplus to use and was disconnected from the GN mainline, but continued as two sidings until the 1960s.To safeguard Rother Viaduct the company agreed to purchase the coal under it on 9th December for £250 per acre.

Messrs. Davies of Rotherham (leased ironstone from the Duke of Newcastle), asked the MSL on 6th May 1856 to provide a siding at Shireoaks. The board replied that it would not provide a siding at its own expense on somebody elses land, but if they would pay for it, they would provide it.

The 1850s saw the expansion of deep shaft collieries towards the east away from Sheffield, with the sinking Orgreave in 1851, followed by Shireoaks 1854–9 and Waleswood in 1858, this colliery was estimated by the MSL in July 1860 to produce 300 to 400 tons of coal per week.

On 26th October the board received a letter from the Midland Railway, objecting to the suggested curve from Woodhouse Mill to MSL mainline by east to south line. The board said that the Midland agreed to west to north curve and Mr Watkins was instructed to talk to Midland Railway over this matter.

A Worksop to Staveley railway was discussed by the MSL board on 26th October 1860, they decided to refer the project to the GN for possible joint construction. Three more deep shaft mines were sunk in the 1860s, with Nunnery 1860–4, North Staveley (later called Aston) 1864 and Kiveton in 1867.

Mr Alderson of Priorwell Brewery approached the MSL with regards to opening refreshment rooms in Worksop station, he proposed to open it in one of the waiting rooms. The board replied on 10th November 1865, that it was not necessary or admissible to open refreshment rooms at Worksop. At the same meeting four improvements were approved, At Sheffield, a problem exists with capacity, direct traffic via Doncaster for South Yorkshire, and add a siding for 80 wagons at an estimated £600. At Darnall, a siding estimated at £1,500 was to be built on spare land. Additional siding on down line, crossing at Shireoaks for around 60 wagons. Estimated to cost £500 (including stop blocks, points and crossings). A siding to hold 150 carriages or wagons to be built at Worksop for £1,500. On 5th January 1866, the board sanctioned the building of an approach road to Worksop station for £30, as well as a new signal box for Woodhouse Junction for an estimated £150. Two months later, the Kiveton Park Colliery approached the MSL for sidings for there new colliery in the process of being sunk.

The company estimated that for £350, they would provide signals, points, etc, half of which, however would be required at preparation work, to be done as necessary. The only work authorised in 1867, was for a siding at Retford, for a maltkiln. For £250, points, crossings, etc, and another £110 for the purchase of land. The Kiveton Park Coal Company had another works at Wingerworth and wished to transport coal there. The Midland Railway called upon the MSL to complete junction at Woodhouse, with new interchange of traffic using agreement dated 16th February 1861. The MSL board minuted :- Until North Staveley Coal Company provide land, there would be a delay of eighteen months. The MSL suggested to the Midland, that they work the trains to a siding on the Midland. The board were told that the North Staveley Coal Company would require a siding, possibly put two together for consideration. It was estimated that it would cost £450 to put in a double junction and a signal box. The connection with the Midland North Staveley Curve came into use 1st January 1870. The last year of the decade, North Staveley colliery (later renamed to Aston), asked for provision of sidings on 28th July and the company expected a large amount of traffic. The agreement was that the colliery built the sidings at there own cost. On 5th November Block telegraph was approved for the section of line from Kiveton Park to Orgreave for £356, plus annual maintenance of £91 2s 10d, this have an advantage of security.Unwins of Darnall enquired if the MSL would put in siding, to which they replied they would consider if a specific request was made.

A deputation from the inhabitants of Handsworth-Woodhouse asked on 8th April 1870, for the removal of Woodhouse Junction Station to a third of a mile westwards, on the grounds that the present station is narrow and inconvenient, and that goods and minerals traffic had so increased, to interfere with passengers. On 20th, Mr Sacrè reported that he had been unable to meet the local people. At a meeting on 2nd June, Mr Sacrè reported to the board that the matter of Woodhouse stations move to the west would have advantages, but there was a question of cost. Six weeks later Mr Sacre reported that they estimated a new station would cost £1,300, the company would move the station westwards, if the people of Handsworth-Woodhouse would pay a portion of move. The board postponed the proposed station on 7th October. Five days earlier Mr Coggan, asked the company to provide a siding (two miles east of Kiveton Park) to a stone quarry, at which he was to erect some lime kilns, and it was estimated that it would provide traffic of £1,000 to £1,500 per year. Two months later, the board were told that :- the gradient on this part of line 1:140, although the line is straight for considerable distance. Mr Coggan to pay whole cost of watchmans wages (signalman?) and cost of block telegraph. The board sanctioned this new siding. The Fence Coal company desired to form connection with Orgreave siding and branch, they were willing to erect the bridge and works at own cost. At Shireoaks the colliery required a much needed two siding connection. The company agreed to the request on the condition, that the company put in the points, and they agreed. Mr Rhodes of Woodthorpe Colliery applied for a siding on Birley branch, half way up the line, a siding for 24 wagons. Mr Rhodes willing to bear all cost and asked to be allowed to do own excavating. The company agreed, provided this is done to satisfaction of Company Engineer. At the last meeting of 1871, the board on 29th December discussed the proposal of a siding at Bole. Lord Middleton offered to give land at £30 per year, subject to legal queries about annual rental. On 29th January 1872, the board heard the request from the agents for Duke of Leeds for a siding at Kiveton Park, to which the company agreed, subject to negotiation. The North Staveley Coal Company agreed with the Midland Railway putting down additional siding,for the conveyance of the traffic, the siding on southside of MSL. Provided at cost of coal company and the Midland, with the exception of the cost of connection between MSL sidings and the sidings to be borne by MSL. The MSL to provide land for siding and the junction between siding and Midland line, to be under Midland control. In 1873, the MSL improved facilities at three stations, Kiveton Park, Woodhouse and Worksop. The first board meeting of held on 22nd January, discussed that a new road should be laid down at Bole level crossing for agricultural in lieu of proposed siding, estimated at £1,200. At Kiveton Park, the board undertook to ask the Duke of Leeds, to purchase about 10 acres of land for sidings and station enlargement. Plans to be prepared for Woodhouse Station, showing how better accommodation can be provided. At Worksop a road to be made to the siding, recently laid down, as authorised in the additional powers act of 1872. The next meeting two weeks later, were told that it was estimated that the new road on the westside of Worksop station would cost £1,479. The question of Kiveton Park was raised at a meeting on 18th July, it was suggested by Col. Rich to substitute level crossing by bridge, estimated at £1,150. The board decided to ask the land owners whether they would contribute to portion of cost. The board next discussed this matter on 19th September, after deferring it at the last meeting.It was suggested to do away with one of the level crossings to the east of the station. Mr Underwood submitted plans for additional siding. The question was referred to Fenton and Brown, who would visit the site and report to the board. They reported back on 3rd October and recommended - Close level crossing at eastern end, divert to level crossing nearer to station. Do away with level crossing would involve large amount of cost. Question of new station accommodation, present inadequate for increasing traffic. Recommend a new station be erected on site a few yards west of present station, and that siding accommodation should be provided at £3,500. Site suggested near to Kiveton Park Colliery, on land owned by Duke of Leeds, and tenders for £4,650 should be made. Tenders for a new station were not received until November, and the contract was awarded on 14th November to Foxton for £1,717. During 1874, some of work discussed the previous year came to fruition. At a meeting on 17th April, the board were told that it was necessary for further accommodation for Woodhouse station. The company would have to negotiate with land-owners for additional land required. At board meeting on 2nd October they were told that the diversion of the road at Kiveton Park would cost £1,260, the board told them to proceed without delay. Retford goods yard required an additional siding and this was approved on 30th October for £1,000.

The additional land for Woodhouse station, was reported as acquired, to the board on 1st January 1875. They instructed the engineer to commence building, and it came into use on 11th October that year. More sidings were authorised through the year, at Beighton for an estimated £7,000 and Worksop for £3,600 on 5th March.

Followed the next month by a siding for Retford Waterworks, and for Messrs Berry at Worksop on 14th May. The Midland branch from Mansfield opened on 1st June, to passenger traffic. The MSL using there powers commenced running 3 trains per weekday from Sheffield to Mansfield calling at all stations, and the Midland commenced with 3 trains to Worksop and Retford using running powers, however the Great Northern did not take kindly to the Midland entering there station at Retford. The MSL provided an engine to take the Midland train to and from Worksop. (For more details see Midland Railway around Nottinghamshire Vol. 1 Chapter 6) The company received on 20th August a memorandum from the inhabitants of Checker House, - that proper accommodation be provided, at present no waiting room, no shelter of any kind and traffic at the station being considerable. Two weeks later, the board were informed that improvements would cost £500, which was larger than traffic can justify. On 17th September, a revised plan was submitted for less expensive accommodation. At the next meeting on 1st October, the expenditure of £150 was approved. The last four years of the decade, saw a great slow down in changes. The Great Northern proposed a new south to west curve at Retford, to be built on a easier curve for faster running. The MSL board discussed the matter on 18th August 1876, and proposed to the GN that the curve become a joint line. It was proposed at a board meeting on 5th January 1877 to light Kiveton Park station with gas. Midland services from Mansfield to Retford were not hauled by Midland engines until 1st July 1878, just over three years after there inception. A public footpath serves as access to Darnall station, and as a result of building a school at Darnall, the company decided to erect a footbridge to cross the line at the station. It was sanctioned on 5th July 1878 at a cost of £350.

The MSL grew in the 1880's with additional sidings being built each year to meet present and new customers demands. On 23rd March 1880, the company sanctioned the building of a shunting siding at Checker House for £650.Messrs Vaughan & Co. asked the MSL on 19th November to extend there siding at Birley Vale Colliery and they approved the work. On the same day the board sanctioned, the much needed footbridge for Worksop station level crossing at £400, however construction was not started the end of September 1881. Nunnery Coal Company approached the MSL to put down extensive sidings, the board were told on 17th December they would cost £6,000, and asked the MSL to contribute towards there construction. The board said, that they would provide the land, but the coal company was to pay for there construction. Work costing £1,800 was approved on 5th May 1882. Shireoaks Coal company required additional sidings for there colliery. On 14th July, the MSL agreed to provide the necessary siding, as well as improving station accommodation for a reasonable price, and tenders were called for. The contract was awarded on 30th November to Messrs Rodley & Sons for £2,630, and was brought into use on 1st February 1884. The residents of Clarborough asked the MSL to build a station, the board minutes of 19th October 1883, stated they could not consider at present. Just over a month later, a request was made for siding accommodation at Checker House, and plans estimated it would cost £960. Work to be carried out, though see if estimate could be reduced. A waiting shed for the down platform of Sturton station was authorised for £100 on 2nd November 1883. The installation of block telegraph between Sheffield No.6 signal box and Woodbourne (Woodburn) Junction, sanctioned on 16th May 1884, for an estimated £190.

After a collision at Retford Goods depot, the company decided after reading the Government Inspectors report, that they needed to improve the goods depot. On 16th May 1884, the MSL sort the help of the Great Northern, with the cost of the improvements. There is no record of a reply, but the MSL agreed to a siding for brake vans to be installed for £150. Messrs Skinner & Halford owners of Waleswood Colliery, asked the company to provide a station, adjacent to the colliery. On 5th September 1884, the board stated, that they were not prepared to recommend erection at present time. Not until 4th September 1885, did the MSL decide to improve Retford goods depot, by providing interlocking signalling at the eastend of the goods yard for £460. Mr Pollitt suggested to the board on 11th December, that an extension of sidings at Claylands (Worksop) was required. The Duke of Newcastle agreed to provide the extension for an estimated £140. In the following year, a shunting siding was approved for Worksop for £160.

The tradepersons of Darnall and Handsworth asked the MSL on 18th March 1887 to provide a goods shed at Darnall. Later that year, the Birley Coal company sank a new pit (Birley East) on the Birley branch, they asked to build a new winning as well as a new branch to the Midland, this was unacceptable to the MSL. Although Treeton Colliery had been sunk in 1877, it was not directly connected to the MSL. On 30th October, the board was informed that two short branches would cost £10,000, and sort powers to construct the branch in November 1887, however this was not successful. In the next year 1888, the MSL forged ahead with alterations and extensions. The Birley Coal Company withdrew there plans 13th April, to connect with Midland Railway. The company after much badgering, agreed to build a footbridge on 11th May, for Worksop station. It was estimated at £310, and tenders were to be got. It opened in November, but the '*Retford Times*' commented that :- "It was in the wrong place, at the eastern end of the station, a proper place is middle of station, say from booking office window". Treeton Colliery reported to the MSL that it agreed to withdraw proposed powers, to pass under the Midland Railway. The MSL board minuted on 12th June, that no difficulty to build, without act of parliament! In November, the company submitted plans for the Derbyshire Lines, of which a Railway No.15 The Waleswood Curve connected the Chesterfield line with the MSL mainline at Waleswood. The line was 1 mile 61 chains long, passing through Norwood cutting 43 to 38 feet deep, and crossed by bridge of 25 feet span over public road. It was proposed to build mostly 1 in 100 and 132. No tunnel was shown on the plans. At the last board meeting of the year on 21st December, it was proposed to lengthen Darnall platform, now that the tickets were collected at that station, and that the old ticket platform at Sheffield was being removed. The board authorised the expenditure of £233 for lengthening. A waiting room on the down platform at Darnall was suggested on 7th January 1889. It was estimated to cost £275, and was approved. The following month Darnall, was again discussed, and it was agreed to improve signalling for £768.

The MSL Act of 26th July 1889 authorised the construction of :-
"Railway No.2. 1 mile 58 chains Parish of Wales, in West Riding, junction with mainline and terminating by junction in Parish of Beighton with Railway No.1".

The board met on 27th July and were told that the wooden bridge at Orgreave was in need of renewal, and to rebuild it with iron was estimated to cost £2,512, they decided look further into it. High Hazels Colliery was sunk in 1889, and the owners Waverley Coal Company, asked the MSL on 16th August to provide sidings to connect with their tramway. The company estimated that these sidings would cost £5,054. A railway was proposed from Shireoaks to Conisborough in November 1889, to pass through possible coal mining area, this railway did not make the statute book. Approval was given by the board on 20th December, for an additional siding at Woodhouse for private owners for £112. At Worksop more sidings for £3,932 and at Bole costing £535.

The next decade heralded great change for the MSL, and the rapid changing circumstances warranted improved passenger and freight handing facilities. The work on improvements continued, with the first meeting of the decade on 22nd January, sanctioned the lengthening and raising of Checker House platform for £284. A siding was authorised for Darnall, estimated at £780, and the board were told that it was much required. The sand quarries of Worksop, brought a large amount of traffic, and to meet this traffic the MSL, provided sand loading accommodation for £155. Two contracts were awarded by the board on 25th April, the first went to Mr Petts tender of £1,000 for the widening of Worksop bridge, and the second was awarded to Albert Ripley's tender of £2,680, to reconstruct wooden viaduct at Orgreave. On 15th August, the board discussed Waleswood station, and Mr Pollitt said he could not decide until proposed line to the north and Conisbrough and to Derbyshire lines, the matter was postponed. Goods yard and sidings at Retford required alterations, and as a result the down loop needed lengthening. It was estimated this work would cost £640 and the company sanctioned it on 27th February 1891. Shunting in Kiveton Park goods caused many problems by blocking the mainline, and the MSL sort to alleviate this on 24th April, by approving the addition of a siding and two short sidings.

To improve locomotive water supplies at Worksop the MSL, decided on 9th October to provide two water columns and parachutes for the station, and the cost was estimated at £295. Rother Vale Coal companies line at Orgreave Colliery branch Junction, caused a lot of problems for the company, there engines being regularly derailed. On 8th January 1892, the board were informed that the Coal company declined to pay for repairs, stating that the Midland Railway, maintains the colliery lines at Treeton Colliery. The MSL under these circumstances, decided that they had no choice but to pay for the repairs, and it was estimated it would cost £1,540 to renew the track.At the same meeting, oscillation of trains crossing Woodhouse Junction had caused a lot of complaints, and the board agreed to the partial renewal. The London North Western Goods station was discussed at a board meeting held on 12th February, at which it was decided that it would like to use it, to alleviate the overcrowded goods stations of the company, however the LNW refused. The MSL took the matter to an arbiter, who said if the MSL wanted to use the goods station, then it would have to extended and the company would have to pay for it. The idea was dropped. Cravens of Darnall wanted to extend there works, but to do this they would have to remove a level crossing, and it was decided that the level crossing would have to replaced by a footbridge costing £485 and that Cravens would have to pay for it. The section of line from Manton Wood to Checker House was the longest without a crossover. The board on 6th May decided to install one at a cost of £200. The Board of Trade inspected the new crossover and approved it.

Eckington (MR) lost its MSL passenger service on 1st June, and the service was transfered to the MSL Derbyshire lines station at Staveley Town. Three days later the service was extended to Chesterfield. At Beighton the connection with the Midland was used only for tranfer goods. The Beighton branch gradually increased its importance as the London extension progressed south.

On 23rd September, the company heard that the Board of Trade would not sanction the opening of the line to Chesterfield, unless Beighton station was provided with a footbridge and covered platform. The MSL estimated that this work should cost £2,357, and tenders were to be obtained.

The Derbyshire lines junction at Waleswood have caused operational problems , and the company decided to resolve this problem, by installing extra running lines between North Staveley Colliery Junction and Waleswood Junction. To relieve the mainline additional accommodation, was to be built at Waleswood Colliery, plus slewing of the mainline. New signalling would be necessary and this would affect the proposed station. The project was estimated to cost £10,994.

Beighton station tenders were received on 6th January 1893, and contract for the platform and station was awarded to Mr Warburton for £1,929, and the footbridge contract was won by Mr J.Gibson for £226 7s. Two weeks later, the board was informed that Mr Warburton had miscalculated the Beighton contract. The board then decided to award the contract for Beighton station to Mr C.Green for £2,530 and the footbridge contract to Mr Ratcliffe for £270.

The Treeton Colliery branch eventually reached the statute book, as part of the MSL Act of 28th March 1893, this authorised the construction of the following railway :-

Railway No.16 55.37 chains long, in the Parish of Treeton, commencing by junction with Orgreave Colliery Branch Railway and terminating by a junction with Treeton Colliey Branch.

Waleswood Curve opened to freight on 17th July 1893, the line had no sidings, but was built with a short tunnel of 66 yards beneath Waleswood road. Passenger services commenced in January 1894.

A suggestion was made at a meeting of the MSL board on 6th April 1894, to build a joint station with the Midland Railway in Sheffield next to Haymarket. The proposal was passed onto Sheffield Corporation for approval.

The water column at Retford was poorly sited for operational use. Should a locomotive and goods train enter the loop to take water, the rear of the train would block the mainline and Thrumpton Level Crossing. The company resolved on 6th April to move the water column and re-erect in a better location for £92. Retford Town Council wrote to the Board of Trade complaining about the excessive length of time that Thrumpton crossing was closed for passage of trains and shunting. On one occasion the gates remained closed for 9 hours 11 minutes. The Board of Trade informed the MSL of the letter and suggested that the company try to reduce these delays.

Locomotives suffered from water shortages around Woodhouse, causing delays and loss of revenue to the MSL. The board decided on 22nd June to erect a 20,000 gallon water tank at Woodhouse. Its installation was estimated at £585. The company received a request for a station at Waleswood, however it was decided that no action to be taken. Kiveton Park Colliery asked for a platform to be constructed. Mr Pollitt informed the board on 13th July, that the land had been purchased and tipping for a rough platform had commenced.

In 1895 the congestion around Sheffield was getting worse, and the MSL tried to alleviate the problem by strengthening the mainline between Sheffield No.6 SB eastwards. A loop line to the north of mainline, plus sidings so that goods trains may be admitted. This work was sanctioned on 22nd February, at an estimated £6,840. At the same time, a water column, parachute and engine pit was to be built at Woodhouse Junction, at an estimated £7,895. The engineer reported a fortnight later, he estimated it would cost £3,894 to build and he did not recommend it. The weather and traffic had taken its toll on Woodhouse Viaduct. Dislodgement of flags had allowed the subsidence of ballast on the viaduct. The board were recommended to strengthen the brickwork, fill-in 4 arches, renovate another 4 arches and fill the span drets with concrete. The board sanctioned the spending of £9,500 for the work. Treeton Colliery asked the MSL build a branch to the colliery. At present coal traffic from this colliery has to pay a toll to the Midland for crossing there lines. At the meeting held on 1st November, the engineer estimated £22,627 to build. The colliery to pay the MSL £200 per year, and save on Midland tolls. Seven years after the MSL were requested to provide a goods shed for Darnall, they decided on 15th November, that a Darnall Goods station with 2 sidings, and a 10 ton crane would be viable, with the stone quarries and nurseries at Handsworth, these would largely pay for its outlay. Due to the heavy nature of the ground, it was estimated it would cost £3,727. The plan for Darnall goods hit a major problem, part of the land had been let for building cottages, so the company had to purchase this section at building land rates, this increased the estimates by £1,900. The board authorised the purchase on 23rd March 1896. Meanwhile contracts for the steelwork for the three bridges, on the Treeton Branch were let to Newton Chambers for £2,094 18s 5d.

Treeton colliery branch opened for traffic on 8th October 1898.

Mr Pollitt agreed at a board meeting on 1st February 1899 to rearrange Worksop station and provide refreshment rooms, as well as removal of water column to another site. Alterations were estimated at £6,483 10s, of which £2,641 was to be paid by the Hotels Committee. The general manager complained to the board that since the London extension, he was not allowed to make improvements to the old main to relieve the congestion. The Waverley Coal company at Darnall (High Hazels Colliery) requested that the GC should install sidings, to which the company agreed, as long the whole cost of £5,361 be borne by the Coal Company. Although the GC would pay £3,191 of the cost, should the coal be only moved by the GC.

Clarborough siding approved on 12th July 1898, was replaced on 12th July 1899, by one at Welham Road because the brickyard closed due to poor quality clay. Rates for goods could be charged at 2d or 3d per ton higher than Retford. The company agreed to spend £1,345 on these sidings. Tenders for Worksop station improvements were not accepted until 6th October. The GC board awarded the contract for £7,751 15s 7d to J.H.Vickers of Nottingham and to be completed within seven months. It was proposed at the same meeting that the GC build a line of 7 miles from Woodhouse Junction to Rotherham, inorder to reduce the clogged lines around Darnall, and engineers were told to prepare plans.

The '*Retford Times*' of 15th December 1899 reported that :-"Great Central Railway commenced renovating and extending station at Worksop, between the station and Carlton Road level crossing. Excavations dug out and proceeding rapidly. The absolutely new building, are externally stone, with two refreshment rooms, one first class 20 feet by 17½ feet, other also fine room of equal dimensions, between them kitchen, modern appliances. Dining room 40 feet by 17½ feet, handsome bay interior by decorating ceiling, paved with wood bricks, walls lined with suitable dados, large cellars, bedroom accommodation, bath and lavatory, placed beyond footbridge under the roof. Awning along this edge, will be in harmony. Remaining buildings reconstructed three waiting rooms, parcels office, booking hall ,offices and couple of rooms for porters and workmen. Stationmasters house little alterations; first class waiting room for ladies, position near covered footbridge, 17½ feet by 15 feet, one detached lavatory and five W.Cs. Large admirably furnished and complete in every detail. Third Class Waiting room, unchanged on same site, lavatories in each case fitted with stoves. First Class Waiting Room, commodious 17½ by 12 feet, similar to that of ladies. Booking Office remain in same place, but separate accommodation for GC and GN, separate Parcels Office. Substantially built awning for traps and carriages. Retford 'up' platform, supplementary to small waiting room, two large waiting waiting rooms, one for ladies and gentlemen plus lavatories.Guards room and lamp room proposed. All existing structures to be pulled down, brick screens to be done awaywith, no projection onto platform as present; platforms to be extended, and lifted by one foot. Present roof made of slate very somber and glass substituted".

The engineer made arrangements with contractors Logan & Hemingway for the building of the four loops to be sanctioned in the 1900 Act. The work to be carried out as soon as the land is obtained. The traffic committee met on 22nd December 1899, and were informed that working drawings were in hand. Mr Rowlinson, explained to the committee, he had asked Logan & Hemingway for a price for the widening, he did not think they would adhere to those quoted. The committee told him, it was desired before inviting tenders in future, he was to submit the position to the Board or Traffic committee.

The Wigan Coal Company were in the process of sinking a pit at Manton Wood near Worksop, and approached the MSL to provide sidings. The board met on 14th July 1897, and were informed that points existed near sidings, it would cost approximately £4,193 to install, and the resulting traffic would be large. At the same time the Coal company wished to have a siding near Checker House, to work the clay. The MSL said it could have the siding for £1,044, but the whole cost would have been borne by the Coal Company. Later that year, complaints reached the board that goods were having to wait for upto seven hours to pass through Sheffield. To improve this serious bottleneck, the company decided that a up and down loop from Sheffield No.6 SB to Woodburn Junction would result in considerable relief, and the board sanctioned the expenditure of £2,885.

Retford Goods No.4 siding was in need of extension, and the GC board sanctioned the expenditure of £167 on 11th February 1898, the sum included the provision of stop blocks. At a meeting on 1st April, Mr Pollitt was told by the Great Central Board to deal with increased traffic at Worksop, particularly the coal traffic off the Mansfield branch. The company authorised the building of shunting neck on up side £213; accommodation for 25 wagon goods yard £1,168; accommodation at west siding for 55 wagons £1,892 and £36 for signals, and giving a total of £3,309. A siding was to provided at Clarborough, to relieve the congested yards at Retford and Worksop. The GC board agreed on 13th July, to the siding and a signal box, estimated at £1,345. The next time the Worksop improvements were discussed, the company found that circumstances had changed. The GC used to sort wagons at collieries, but due to limited space available, shunting had to be done on the main line. The Midland Railway terminated this method of working.This caused the company severe problems, now that the traffic being brought to Worksop. Private sidings were being used to hold wagons, and this interferred with there use. To build a loop line and additional sidings would cost £9.241.

Part of the Great Central Act of 12th August 1898, enabled the company to construct the Manton Wood branch :-

"A railway No.4 22 chains or thereabouts in length, wholly in the Parish of Worksop, commencing by junction with the companies railway, at a point 26 chains or thereabouts along that railway in a south easterly direction from the centre of the bridge carrying the railway over the public road leading from Worksop to Retford, and terminating in a south westerly side of Elkesley Road at a point in that road 14 chains or thereabouts measured in a south easterly direction along that road from the junction of that road with said road leading from Worksop to Retford".

The Great Central Act of 25th June 1900, enabled the company to widen the following sections of line :-

"A widening No.2 of mainline railway of the company on the northside thereof, wholly in West Riding, in parish of Attercliffe cum Darnall at a point on that railway distance of 3 chains in a westerly direction from Darnall Junction Signal Box and terminate in parish of Handsworth in the Urban District of Handsworth at the westend of Darnall Tunnel on that railway.

A widening No.3 of the mainline railway of the company the northside thereof wholly in Parish of Handsworth, at a point on that railway, distant 9 chains or thereabouts measured along the railway in a easterly direction from east end of Darnall Tunnel and terminating in the said Parish of Handsworth at a point on that railway, distant 5 chains or thereabouts along that railway in a northerly direction from the footbridge at Woodhouse Station.

For protection of Corporation of Sheffield in place of footpath crossing of the mainline of the company on the level in the Parish of Attercliffe cum Darnall, between occupation road marked 2 on plans of Widening No.2. and Darnall Railway Carriage Wagon Works. The company shall erect a footbridge and provide and fix adequate lamps with supply of gas for efficient lighting of the same, such lamps to be cleansed, lighted and extinguished by the servants of the Corporation and the company shall defray at half yearly periods to the corporation the reasonable expenses so incurred by the company.

2. The company shall also provide and fix adequate lamps in and shall efficiently light with gas at all times the subway crossing underneath the mainline at the western extremity of Darnall Station and shall be cleansed etc.

3.The company shall widen the existing bridge carrying the mainline over Owlergreave Road to a full width of 40 feet and in consideration thereof the Corporation shall convey to the company free of charge all there interest in two pieces of land abutting on both sides of the railway numbered 22 and 23".

Estimates for widening authorised in the 1900 Act were given to the board at there meeting held on 26th June. Down loop between Darnall and westend of tunnel £15,664; Up loop from eastend of tunnel to Woodhouse £15,745; Down loop Woodhouse to eastend of tunnel £20,977 and Down loop from Darnall Tunnel to Darnall Junction £17,061. At the same time it was estimated that a siding at Retford would cost £6,899.

The *'Retford Times'* of 21st September 1900, reported the opening of Welham Sidings :- "For about 15 years the celery and other growers from Clarborough, Welham and Clayworth struggled to get there own sidings, as Retford was a long way to go. Construction commenced in April, and permission had to be obtained from Nottinghamshire County Council, to lay temporary treadle bridge to take spur wagon to other side.It holds 35 wagons on two lines, a cart road 30 feet wide inclined at 1:20, plus weighing office and machine". Soon after the people of Welham approached the GC to build a station, the company decided that there was not enough people to warrant a station.

The traffic committee received regular updates on the progress of work on the Darnall to Woodhouse loops, and the engineer reported on 4th October 1900, that he was satisfied with the progress being made.

The traffic committee recommended on 26th October, that they construct a footpath for £180 from Darnall station to Handsworth bridge. On 22nd May 1901, the engineer, said that the Darnall - Woodhouse loops will be able to passed to traffic in about a months time, this did not happen. The report given on 19th July, indicated the engineer was not happy with the contractors, it is minuted he made strong representations to have one of the loops open as soon as possible, they promised one loop to be available in August. It is recorded at a meeting on 5th September, that the contractors handed over one loop to the traffic department. Minutes dated 18th July 1902, state that the last loop had been opened.

Construction of the refreshment and alteration to the old station at Worksop were completed by May 1901. The *'Retford Times'* of 3rd May, said that :- "It would have opened on 1st May, but Magistrates refused it a licence. Dining room accomodation for 8 tables and able to accomodate 60 guests. Refreshment bars of uniform pitch pine. Under management of Mr F.Clark formerly of Victoria Hotel".

On 5th September 1901, the Traffic Committee decided to convert a siding into a down loop, and this would enable goods trains to enter Park Goods Depot without blocking the mainline. As a result of this decision the LNW goods branch would subject to blocking, the GC agreed with the LNW to provide a shunting neck. It was estimated this would cost £440 (£300 for permanent way and £140 for signalling).

Rother Vale Coal company required an additional siding at Orgreave, estimated at £4,135. The traffic committee minuted on 20th March 1903, that until something was done, the coal company would divert the traffic, however when completed the company would get increased tonnage. Messrs A.G.Burnett wanted to have an office in Park Goods, the company agreed to this request, it was costed at £237, and they were to pay 10% plus £3 10s per year. The contract for the Messrs Burnetts office was awarded on 27th September to Messrs Chadwick who tendered £159, the work to be completed within six months. To meet demand the company on 22nd October, authorised the building of 2 more cattle pens for Worksop, estimated at £104.

The supply of 36 lamps for Worksop Yard was let to Giddings & Dacre for £108 on 24th May 1905, and the major siding works contract was not awarded until 26th July. Chas Baker of Sheffield won the contract with a tender of £7,875 19s 10d. The *'Retford Times'* on 1st September described the following :- "New concentration sidings for Worksop, become centre for distribution for Mansfield, Dinnington, Laughton, Shireoaks and Manton, the new sidings formed of what is called a gridiron principle, along eachside of existing line. Economy of working is apparently one of its objects, engine shunts its load into loop, a falling gradient 1:80 allows vans or wagons to descend to directed position and the engine is thus able to save time".

Darnall Tunnel to Sheffield Victoria was a very congested section of line, and the board approved the construction of a down loop line from the westend of the tunnel to Woodburn Junction skirting around the southside of Darnall station. The contract was awarded on 20th April 1906 to Chas. Baker of Sheffield, who tendered £2,325 13s 11d, less than the engineers estimate of £3,422.

Handsworth Colliery was being sunk by the Nunnery Coal Company in 1907, and the coal company approached the GC to provide sidings to connect with the colliery tramway. The tramway would cross the mainline by a bridge. The GC board awarded the contract to Chas. Baker on 23rd November for £1,454 4s 7d. The coal company to pay the whole cost, less 10%.

Since 1884, Skinner & Halford owners of Waleswood Colliery had pressed for a passenger station, and they eventually proved successful. The *'Worksop Guardian'* announced on 23rd May 1905, that the GC said it would open a station at Waleswood, but the plans have been lost. Roper & Son of Sheffield, submitted a tender of £2,447 and was accepted by the company. Waleswood station opened on 1st July 1907, and the *'Retford Times'* of 5th July described the opening :- "Flags enthusiastic flying, well equipped commodious, excellent train service, creation of station due to efforts of J.H.Ashton Managing Director of Waleswood Coal Company. Rapidly growing village of 100 houses. Nearest stations Beighton 2 miles and Kiveton Park 3 miles. Timber structure, no buildings or very substantial kind, these won't be of a surprise if you know the district".

The board met on 27th January 1911, and decided after being told that Nunnery Coal Company workings were approaching Darnall Tunnel and wanted to mine the coal under it. Engineer says that tunnel structure will not allow any disturbance, and to purchase the coal beneath the tunnel would cost £18,000. Save £265 per year by closing Tunnel Signal Box and another £320 in tunnel maintenance. Delay of traffic would be greatly reduced. To open out Darnall tunnel would cost £41,500. It was decided to remove the tunnel and invite tenders. Goods depots in Sheffield had reached breaking point. At a meeting held on 7th April, it was decided a new depot would be built at Blast Lane. It would take general traffic and the overflow from Park and Bridgehouse goods.

Contracts were awarded on 16th June, for Blast Lane goods and the opening out of Darnall tunnel, to Pattinson & Son £6,654 6s 6d and Pauling & Co £27,330 3s 3d respectively. The latter was subject to engineers approval. It was not until 6th October, that the engineer recommended a contractor, and it was not Pauling & Co, but Messrs R. McAlpine who tendered a higher quote of £31,396 14s 10d. The board agreed with the engineer.

The bridge carrying the GC over the Midland, was in need of renewal. Messrs W.Richard & Son tendered £852 17s 6d and the board awarded this contract on 7th July 1911.

The two signal boxes East and West Trent were built on stilts either side of the viaduct controlling the Great Northern and Great Eastern Joint line junctions, they were merged into a single box on the eastside of the bridge. The new box was brought into use on 25th April 1913.

A report was given to the board on 24th October, saying that 95% of Darnall tunnel removed, side walls demolished, slow due to large quantity of rock, bridge throughout completed, permanent way between Orgreave and Woodhouse now being laid. Slip on downside necessitated single line working, normal working resumed on 11th. Excessive rain 1½ inches in 24 hours. The next report was received on 20th November, it showed that the permanent way, on the whole of loop Orgreave to Woodhouse proceeded with. Plans for extension of loop to Woodhouse East Junction is in hand. This was not proceeded with in the Great Centrals time. Orgreave to Woodhouse up loop was brought into use on 4th January 1914.

It was suggested on 9th January, that the waiting room at Beighton should be converted into a booking office. The work to be carried out by Abbott & Bannister for £32 18s 6d. Manton viaduct had suffered subsidence caused by mining, the company gave the contract to Abbott & Bannister to supply 24 timber centres, complete and delivery to Worksop for £420. Minutes record that the following were at work on Darnall widening -; 3

steam cranes; 1 steam navvy; 14 locomotives; 1 portable; 32 wagons and 110 men. Two months later on 6th March, work on Darnall tunnel was recorded as complete. Cementing of Manton viaduct was reported on 24th May. The widening of the Great Central lines over the Midland were finished on 18th November. Of the works authorised before the First World War broke out, Blast Lane Goods was the last to be completed on 6th January 1915.

The paving of Park Goods for £2,220, was approved on 1st February 1916 and was the only major work to be approved through the First World War.

After the Great War, work commenced on catching up with maintenance, paint contracts were let in 1919 covering Sheffield to Sturton, and reports showed only 90% was complete, the question of the renewal of Worksop Station roof was discussed at a board meeting on 5th March 1920, and tenders were invited. The contract was awarded to T.Carr on 16th March for £7,771. The work involved the removal of the roof and replacement by two glass covered awnings. Work commenced immediately, with a report given on 28th July stated the steelwork was on site. On 24th November, a portion of the roof had been taken down. The next report dated 2nd February 1921 20% completed. Three months later the reroofing was 90% complete and by 21st April, it was sufficiently complete to sign a painting contract. This was completed by 27th June.

A workmans service operated from Sheffield Victoria to Orgreave Colliery, it commenced after July 1900. The 1922 timetable was as follows :-

Weekdays				SX	SX	Sunday
	am	am	pm	pm	pm	pm
SHEFFIELD	520	535	110	9 2	1012	1035
DARNALL	526	541	116	9 8	1019	1042
Orgreave Coll. Sid	530	545	120	912	1022	1045

Weekdays				SX	SX
	am	pm	pm	pm	pm
Orgreave Coll. Sid	624	218	240	947	1045
DARNALL	627	221	243	953	1048
SHEFFIELD	634	228	250	958	1055

The LNER took over from 1st January 1923, and immediately undertook to improve facilities. The Traffic committee met on 4th October and were told that Orgreave sidings at present hold 120 wagons, which was inadequate, and engines were delayed, resulting in a loss of traffic. It was estimated that at a cost of £4,600 an additional siding to accommodate 162 wagons could be built, and it would save £1,834 per year in engines. It would increase traffic to £18,049 per year. The extension of the up and down goods loops from Woodhouse station to Woodhouse East Jn was approved on 13th December at an estimated £46,544. The loops would pass round the outside of Woodhouse station and join at the East Junction, necessitating a rearranged junction as well as an additional siding. The company would save £5,550 per year in reduced delays. It was proposed at a Traffic committee meeting on 21st February 1924, the up goods loop should be extended between Darnall East and Woodburn Junction. This would entail rebuilding Darnall Station, rearrangement of up and down mainline, and down loop, as well a shunting neck to hold 15 wagons to connect to third goods siding. Estimated to cost £35,991 (Loop £19,110; Darnall Station £12,574; Shunting Neck £3,557 including £1,000 for land and £750 for an additional crossover), but would save £3,931 in annual operating costs.

An error in calculation of the Woodhouse goods loops was found and reported to the Traffic and Works Committees on 31st July 1924, it increased the original estimates by £10,419 to £56,963. The committees instructed that it be carried out as soon as expedient. The contract for extending Darnall up loop was awarded to Charles Fletcher & Co on 2nd October, for £34,908 5s 0d. Chadwick & Co won the contract to build Darnall Station (New station buildings, platform and roofing) for £10,419 on 8th January 1925. The Woodhouse loop extension required parliamentary powers to divert a footpath and part of a road. These powers were included as part of the LNER Act of 31st July 1925 :-

"Stop up and discontinue use of footpath. North West of Woodhouse station 1½ chains North East and 1 chain South West of said railway, divert the same over Company footbridge, extend footbridge and make additional opening therewith. Divert in North Eastern direction so much of road known as Soap House Lane as lies between Furnace Lane and at 36½ chains South East, therefore stop up and discontinue use of Soap House Lane between said points".

The LNER submitted plans for a Gedling to Retford railway in November 1924, for the 1925 parliamentary session. At Retford, it crossed the GN mainline by a bridge span of 58 feet and joined the GC line west of Gringley Road level crossing at Milepost 65. A spur left the Worksop to Retford line at Milepost 63½ and extended 59 chains to join the Gedling to Retford railway at The Laurels. Thus the LNER would have a Retford avoider. However the plans were not successful and were withdrawn in favour of a joint railway with the LMS, which joined the GC near Checker House. (See Midland Railway around Nottinghamshire Volume 2).

Later that year on 26th November, the LNER proposed to close Kiveton Park West Colliery Signal Box, which would initially cost £940 to close and transfer its functions to the East Signal Box. High Hazels Colliery complex closed in September, but traffic was still exchanged with Handsworth Colliery. The committee meetings held on 25th March 1926, it was reported that the extension of the up and down goods was delayed, because of a delay in agreement with Sheffield Corporation. It was approved that Retford East Signal Box would close, for an estimated £2,487, but saving £385 per year (3 signalmans wages £487 less £102 maintenance). The Traffic committee on 1st July 1927, approved the purchase of 157 square yards of railway land at Worksop by Turner & Sons for £40, but subject to firm option to acquire in next 12 months, 6 acres 2 rods and 9 perch of land, required for the extension of the marshalling yard for £2,500 (land part of Turners Sand workings). They were also told, that the firm is prepared to excavate sand and top soil within 6 months for the LNER, but decline to pay for privilege. Should the company undertake the work, it was estimated it would cost £16,000.

During 1924-6, Shireoaks Coal Company was in the process of sinking a colliery (Harry Crofts or Shireoaks No.3) just west of Brancliffe Junction. A Traffic committee meeting on 2nd June 1927, were told that a temporary connection had been put in 1922, and it was expected to produce 1,020 tons per day from August 1928. The coal company asked for the LNER to provide sidings and to increase them as the colliery developed. The LNER agreed to construct a connection to up and down lines, install new Signal Box and signalling, and it was estimated at £4,538 plus annual maintenance of £163. The LNER to pay 50% of the costs, but it was estimated that the revenue would be £40,000 per year. Harry Crofts colliery never reached its expected capacity and closed on 10th August 1929, due to ventilation problems.

Handsworth Colliery wanted to build 25 coke ovens on or near the site of High Hazels colliery. The traffic committee discussed the request for sidings on 31st May 1928. Present sidings hold 181 wagons and will be inadequate. It was expected that the inward traffic would consist of 360 wagons or trucks and the outward traffic would be 130,000 tons of coal; 100,000 tons of coke; 6,000 tons of tar; 2,400 tons of Sulphuric Acid, and 48,000 tons of Crude Benzol. Two additional loaded sidings at an estimated cost of £4,015 would be required.

1. The coal company to provide the land and entire cost of the scheme.
2. LNER to refund estimated £2,362 of permanent way and materials at 10% rebate of portion of receipts of traffic to and fro.
3. Two additional sidings, maintenance by LNER, expense to be borne, cost of labour paid by coal company and cost of material to be paid by LNER. It was estimated that it would generate £5,730 per year revenue.

The Works committee meeting of 27th July 1928, found that the estimated cost of mechanical signals for Harry Croft Colliery would be £2,017. At the next meeting, the tender for installation and supply was awarded to British Power Railway Signal Company for £1,994 9s 7d. Wales Parish Council had repeatedly applied to the GC and LNER for a station for Wales. A Traffic Committee meeting on 29th November, were informed that Wales Parish had a population of 5,000 and 452 new houses had been in the last year, and the nearest station was 1½ miles distant at Kiveton Park. The Kiveton Park Coal Company had undertaken a verbal agreement to the LNER, that if receipts from the new station were not sufficient to cover outlay and manufacture, they will divert to LNER coal traffic now passing by the competing route (LMSR). As soon as the station opens, they will transfer 50% of the pit prop and mining timber traffic. The whole of which cornered by the LMSR. The committee were told that this gross £1,500 per year. The committee recommended a 300 feet long by 8 feet 6 inch wide platform, and necessary office accommodation, waiting room and conveniences. Located in the centre of Wales village, 1 mile 15 chains from Kiveton Park Station and 1 mile 48 chains from Waleswood Station. The station estimated to cost £2,057 to build. The Works Committee asked for tenders to be invited. The Traffic Committee also were told of the annual operating costs of the new station, it would come under the control of the stationmaster at Kiveton Park. The staff would consist of one 5th class clerk at £200 maximum, one porter including uniform and holiday relief £114. Stores, lighting and maintenance of £35 per year. Estimated receipts to Sheffield of £650, plus pit props to Kiveton Park Colliery. A profit of £750 per year was expected.

Brookhouse Colliery was the last colliery to be sunk next to this line, sinking began in 1929.The LNER traffic committee met on 6th June, and were told that the colliery output was estimated at 250,000 tons per year. Building of sidings to serve it, and were estimated to cost £9,486. The annual maintenance costs of £45. Aston (North Staveley) colliery was expected to close and be replaced by Brookhouse colliery.

Kiveton Bridge station opened to passengers and parcels on 8th July 1929. The '*Worksop Guardian*' of 12th July, reported "First train 9 a.m. from Sheffield and the station was informally opened. After 30 years of agitation a station in centre of Kiveton Park. There was a large crowd of spectators for 9 a.m. arrival from Sheffield pulled in, and a loud cheer as train steamed out. Lots of competition for the first ticket. District Manager Mr E.H.Cause (Lincoln) presented the first ticket to Mr E.P.Hall and second to Mr H.S.Turner, both of Kiveton Park Coal Company. Also present District Inspector A.Kitchen (Retford). Some dissatisfaction at no trains running before 9 a.m. and particular 8.28 a.m. not stopping to pick up large number of secondary scholars and business people for Sheffield. After 9 a.m. there is a very convenient and frequent service. For present no Sunday service".

The 1930s proved to be very uncertain, with its major traffic from the coal industry,was subject to a prolonged recession, however modernisation continued.

A sand drag and facing points next to starting signal of Retford South Signal box on the GC avoider, were approved on 20th February 1930. The locomotive shed at Neepsend was in need of a new roof at a cost of £15,000, as well as only be able to hold 42 engines. The Traffic and Works committee of 1st May 1930 decided to build a new depot at Darnall, and costing an estimated £238,186 and be able to hold 112 (For more details see Chapter 5). The scheme was abandoned on 25th September, as the terms of the Government assisted works programme was not as large as the LNER expected. The proposed depot site, was owned by Church Burgesses Trust (2 acres, 2 rods and 6 perch of land), at an estimated value of £1,015. Wire Rope Co. land to be exchanged with similar land in the area. The Corporation wanted the part of the site for allotments an area of 8 acres, 3 rods and 20 perch, worth £2,750. Even though the Darnall scheme was abandoned they were recommended to purchase the land for the future. The board authorised £7,700 for purchase of all the land required. The new sidings for Handsworth Coking Plant were completed in November 1928. By 23rd November 1930 the

LNER Traffic committee were informed that the actual cost borne by the company was £1,836 and it has been refunded. Extra work of £1,886 of which £927 was paid by the LNER. During 1929, 225,806 tons of traffic were transported out, and produced £6,500 of revenue. It is necessary to extend accommodation at an estimated cost of £325. The coal company to lay out at there own cost and the LNER to pay £602. An additional siding for empty wagons is required for an estimated £142, and this would save on shunting. The plans for Brookhouse colliery were changed at a Traffic Committee meeting on 27th November 1930, the sidings were estimated to cost of £12,305, less £932 for recoverable materials. The footpath over the line would need to be extended at a cost of £500, however the coal company tried to have it blocked up (they were unsuccessful). The new scheme would have delivery and empty sidings, and this save £420 in operating costs against operating costs of £1,887. It was estimated that the additional coal and coke would generate £9,835 in revenue.

The economic depression of 1931-1936 resulted in stagnation of traffic and projects being deferred.Aston Colliery ceased production in April 1931, and the new colliery at Brookhouse was to take over. The only casualty for passengers, was the withdrawal of services from Checker House from 14th September 1931, after the end of the summer timetable. The area now being served by East Midland Motor Services and Sheffield Joint Omnibus both with a LNER shareholding. Brookhouse Colliery, now technically part of Beighton (Birley) colliery, was stopped from producing any coal because of the Coal Act strictly limited production.

As part of the MSW scheme, a document dated 30th September 1935, said that additional sidings would be required at Woodhouse, for interchange of traction. It was estimated these sidings would cost £45,770. The eastern limit of electrification would be 46 miles 7 chains and 45 links from Manchester London Road. The The Traffic committee were told on 25th October 1935, that work at North Staveley (Brookhouse) colliery had been held in abeyance and no prospect of the colliery being opened, they rescinded minutes authorising work.

From 1936, traffic began to pickup, the weigh bridge at Thrumpton Goods, now over 85 years old, needed replacing, because although it could nominally weigh 8 tons it had to be restricted to 3½ tons. A new weighbridge to weigh upto 20 tons was authorised for £351, by the Traffic Committee meeting on 9th January 1936. Later that year the Works committee on 29th October, discussed the Manchester, Sheffield and Wath electrification scheme which extend as far east as Woodhouse. They decided to defer the £2,568,945 scheme, to see how much the Government assisted works may possibly provide. Shell Mex BP were dissatisfied with the present distribution facilities provided by the LMS in Sheffield. The Traffic Committee on 28th January 1937, were informed that the transfer to the LNER would net £6,000 revenue per year. They decided that if Shell Mex BP would undertake to distribute all the oil for the Sheffield district, from there depot, the LNER would provide facilities at Woodburn Road Siding. The engineers would have to vacate the road they use, and the company would have to widen the gate in the fence, and it would cost £451. A rental of £36 per year, plus maintenance charges of £6.

The Manchester, Sheffield and Wath (MSW) electrification scheme got off the ground, a new exchange yard was to be built at Woodhouse (actually Rotherwood). As part of this the Works committee on 23rd June 1938, authorised the provision and construction of engine turntable, foundations for engine pit, water crane, foundations and substructure for water tank. This project was Government assisted and was estimated to cost £2,518, but Fletcher & Co. tendered £2,170 6s 2d, and were awarded the contract. The next Traffic and Works Committees meetings on 21st July, authorised various contracts for signalling and the new steam and electric depots at Darnall. A new permanent way connection from the electric and steam to down loop between Darnall West and Woodburn Junction. It was to be controlled by switches from ground frame, but this was decided as unsatisfactory, and now to be operated from Darnall West at an additional cost of £960, plus £26 in annual expenses. Ordsall Lane level crossing was not protected by signals, and the Ministry of Transport, said it was desirable that up and down signals should be interlocked with the gate. The Traffic Committee authorised the expenditure of £302, at a meeting on 29th August.

The LNER Works committee first meeting of 1939 on 5th January, authorised the rebuilding of an overbridge at Worksop be let to Motherwell Bridge and Engineering Company for £6,179 9s 6d, this was well over the estimate of £5,000. Another contract for the MSW scheme was approved on 16th February, for Darnall Flyover to G.A.Pillatt & Son Ltd for £9,201 6d 9d, which was lower that the £11,000 estimated. This bridge would ease congestion for steam and electric engines gaining access to the depots. At the last meeting before the outbreak of war, two contracts for conversion of track circuits from DC to AC and the supply and installation of signal and signalling equipment at Orgreave and Rotherwood was awarded to Siemens & General Electric Railway Company for £24,300 and at £17,173 respectively. The Ministry of Aircraft Production asked the LNER to provide a siding at Worksop at an estimated cost of £355, the Works Committee on 27th May 1940, approved the work as the Government were to pay for it. From November 1940 to September 1945, no LNER committees appear to have met, most of there work was conducted by various committees at the Railway Executive. During the war Darnall Steam Shed was brought into use (See Chapter 5 for details) as well as Rotherwood Yard, which appears on a LNER track plan in April 1944 as Roundwood Yard. The yard was composed of 4 up and down sidings for the exchange of motive power. It appears that most of the work was undertaken by LNER labour and was in use just before the war.

The Birley branch lost most of its traffic with the closure of Birley East and Birley Vale collieries in 1945, and converted into a training centre.

After the war, the LNER committees recommenced on 29th September 1945, they immediately authorised the installation of a 100 lever frame for Woodburn Junction Signal Box and alterations to the permanent way to and from Darnall Motive Power Depot, at an estimated cost of £3,966. Facilities at High Hazels Yard were improved by the Traffic Committees decision on 21st February 1946, for the erection of brick building, shunter and lengthmens cabins, at an estimated cost of £1,020. One week later the same committee, approved the construction of toilets for Whisker Hill Sidings, for an estimated £617. The decision concerning the provision of locking signals to Ordsall Lane crossing was rescinded on 25th July 1946. The last year of the LNER existence, the Works committee on 27th March 1947, agreed to purchase an electric flash butt welding equipment for Beighton depot at an estimated £41,872. The committees on 3rd July 1947, were informed that the MSW scheme would be proceeded with as a satisfactory arrangement with the Government had been concluded. The estimated cost had risen from £3,189,895 to £6,200,361. The last project to be undertaken by the company, was given to go a head on 30th October, for a District Engineers Workshops at Woodburn Junction.

The nationalisation of the railways from 1st January 1948, brought little change with the line now under control of the Eastern Region. Waleswood Colliery ceased production in May 1948. Interchange sidings at Rotherwood required an additional up loop, to meet altered arrangements. The Eastern Region could not undertake the work until after Stage 3, and on 12th October 1951, a contract for £8,317 10s was awarded to Wellerman Bros. of Sheffield to build this loop.

Fire destroyed the interior of Waleswood station booking office on 24th May 1953, and caused £2,000 of damage. The fire raged so fiercely that traffic was delayed for sometime. Nunnery Colliery ceased production in August, but the Coke Works continued until at least 1956. Passenger and freight services were withdrawn from Beighton on 11th November 1954.

Stage 4 of the Woodhead electrification scheme from Sheffield Victoria to Rotherwood Exchange Sidings was energised on 3rd January 1955, thus completing the 1935 project. Freight could now be electrically hauled across the Pennines for exchange to steam. Waleswood Station closed to passengers from 7th March 1955, the last service being on 5th March as there was no Sunday service.The coking plant closed in 1957, some 9 years after the colliery ceased production. Checker House goods facilities were withdrawn from 1st October 1956.

The winter timetable of 1958 starting on 15th September saw the diversion of '*The Master Cutler*' from Marylebone into Kings Cross calling only at Retford. The service was upgraded to Pullman and a fillin turn ran to Sheffield late morning and returned late afternoon and this was known as the '*Sheffield Pullman*'. From the start the premier service was entrusted to diesels.

Sturton closed to passengers from 2nd February 1959, the goods sidings became an unstaffed public siding. Goods traffic was reinstated to Checker House from 31st October 1960.

Woodhouse Viaduct became unstable from about 15th June 1960, and all locomotive hauled trains were diverted via Killamarsh Junction, where a reverse was necessary, and diesel multiple units crossed over at 5 mph. The Winter timetable of 1960 saw the recasting of the Sheffield Victoria to Manchester Piccadilly service with the consequence that connections to the east involved long waits. The summer of 1961 brought the end of scheduled passenger services over Waleswood curve. It was only served by one train each way, Saturdays only from 1st July until 26th August, the 7.36 a.m. Chesterfield Central to Skegness and 1.47 p.m. return from 1st July until 26th August.

The early 1960s brought the construction of 2 coalfired power stations in the at West Burton and Cottam, it was to be supplied by coal from the collieries in the Notts, Derbys and Yorks coalfield. A daily traffic of 21 trains would have to cross the mainline at Retford, the problem of conflict of traffic was to be solved by the building of a dive–under. Blast Lane Goods was the last depot to built in Sheffield and was the first to close on 6th November 1961. However a siding was built to the newly opened Parkway Wholesale Market in 1961. Handsworth coking plant ceased production in 1962 and the railway lost a significant source of traffic.

Construction work was sanctioned in the "The British Transport Commission Act of 1st August 1962, authorised the following -:

Work No.14 A railway (1 mile 558 yards in length) commencing in parish of Babworth in Rural District of East Retford by junction with railway between Sheffield and Grimsby, at a point 1,088 yards west of bridge carrying Ordsall Road over the said railway and terminating in the Borough of East Retford by a junction with said railway at a point 45 yards east of the point where Thrumpton Lane is crossed on the level by that railway.

Work No.15 A railway (721 yards in length) wholly in the Borough of East Retford by a junction with with Work No.14 at a point 610 yards west of the road carrying Ordsall Lane over the railway, between Sheffield and Grimsby and terminating by a junction with Whisker Hill Curve at a point 95 yards North East of the said bridge.

Subject to provision of 1845 Act the company may construct the bridge, Act of 1853 with crossing of roads on the level. The company in connection with construction work No.14 carry the same by 2 lines of railway across and on the road of Thrumpton Lane numbered in the deposited plans 12 in Borough of East Retford".

Work commenced on the dive–under in November 1963, with contracts let mainly to Henry Boot of Sheffield, the first work involved a temporary northspan for Thrumpton Lane footbridge, and this enabled the track to be slewed to the north and demolition of the southern half of the River Idle bridge. Excavation of the Bunter Sandstone had to be undertaken without spilling off the railway land.Ordsall Road and West Carr Road bridges were diverted, to allow work on foundations. The east coast mainline was placed onto a temporary bridge during January 1964, to allow excavation of sandstone. Parcels lift and low level platforms were completed in May, and the permanent bridge was rolled into place on 14th June. The low level platform footbridge was inbuilt to the east coast mainline diveunder bridge. Flyover deck of Ordsall Road bridge was put into position later that month, creating a 3 tier bridge. West Carr Road bridge reopened to traffic in August. In November 1964 ballast was laid down in the dive under and Thrumpton crossing was lowered.

Local goods facilities were beginning to be rationalised in the move towards bulk load traffic. In 1963, four depots were closed, Park (Sheffield) closed on 1st June, and Rushey Public Siding 4th on November, and followed by Checker House and Shireoaks from 9th December. Sturton survived until 6th June 1964 and Kiveton Park closed on 14th June 1965.

West Burton Power signal box opened in Spring 1965, in readiness for the opening of the power station. The

Summer timetable of 1965 was due to commence on 14th June, and that weekend was chosen to complete the diveunder. On 12/13th, all East Coast services were diverted and cross country trains terminated at Gainsborough and Worksop. That weekend the engineers had to remove the flat crossing, connect the westend of diveunder, and demolition of Whisker Hill Signal Box to enable down loop to connect onto the Retford High level line. Signal boxes at Welham Road, Gringley Road, Ordsall Crossing, Whisker Hill, Rushey Crossing and Checker House were closed and replaced by Thrumpton Crossing Signal Box fitted out as a mini power box. Colour light signalling enabled a 5 minute headway at 35 m.p.h. from Clarborough Junction to Manton Wood. Gringley Road level crossing became electrically operated from Thrumpton Signal Box. A refuge siding was built in diveunder on the downside, west of the Low Level platforms.

All cross country services commenced using the diveunder from 14th June, although the work was still not complete. The up platform shelter was not ready, and two footbridges over the River Idle and Tunnel Road were not expected to be complete until the Autumn. As part of the scheme a down loop was constructed to the east of Rushey Crossing and the new diveunder junction, and it came into use before the diveunder opened. As a result of the diveunder, Whisker Hill Junction into Retford High Level was reduced to two passenger trains per day and they were the "Master Cutler" and "Sheffield Pullman".

The closure of Retford GN shed saw the stabling of diesel locomotives at Worksop (usually in old cattle dock). The shunters being provided by Shirebrook 41J and the Brush Type 2s, English Electric Type 3s and Brush Type 4s from Immingham 40B, Tinsley 41A, and Wath 41C.

It was proposed as part of rationalisation in early 1965 to concentrate all possible traffic on Sheffield Midland as being a more central station. A connecting spur was installed between the Midland Nunnery Colliery branch and the Great Central a quarter of a mile east of Sheffield Victoria station. From 4th October all but a handful of trains from Retford, Lincoln and Cleethorpes were transferred, making an inconvenient walk for Manchester passengers.

The line between Woodhouse Junction and Beighton passengers service was reduced greatly with the demise of the London extension, local services ceased after 4th March 1963, and only four trains survived to closure on 5th September 1966. The last service was as follows :- Sheffield Victoria southbound 0009 (ex 2220 York) – Swindon; 0045 (ex 2305 Manchester Central) – Marylebone; 1137 (1008 ex York) – Poole and 2055 (ex 1840 York) – Swindon, and northbound arriving at Sheffield Victoria 1800 (1030 Poole – York); 0057 (1920 Swindon – Sheffield); 0141 SO 0151 SX (2245 SO 2155 SX Marylebone – Manchester Central) and 0210 SX (2150) 0243 SO (2225) Swindon – York. Upon closure to passengers the line was retained for freight use, as well as diverted passenger trains.

Worksop was decided to be the local maintenance depot for the Merry–Go–Round wagons, and a three line shed was started to be built in Autumn 1966 and was completed in time for MGR scheme to commence.

The connecting line from Killamarsh Junction to Waleswood Junction was closed to all traffic from 8th January 1967, with the former LD&EC line traffic diverted over the new Langwith Spur and the Arkwright Colliery traffic could easily dealt with by running round at Woodhouse East Junction. West Burton Power Station came on stream this year, a little later than expected, and the Merry–Go–Round (MGR) coal trains started to run. Brush Type 4 locomotives and 30 wagons ran straight from the colliery to the power station. Handsworth Colliery ceased to produce coal from October 1967, and Handsworth sidings lost its last source of traffic.

The 'Master Cutler' and 'Sheffield Pullman' services were withdrawn from 7th October 1968, as the Pullman did not fit into new Inter–City image. The last day service of two round trips was hauled by Brush Type 4 1525 (41A). Due to a derailment at Peterborough, it arrived at Retford 37 minutes late and missed the connection to Worksop, and so on its last leg the Master Cutler called additionally. Probably a first and definitely the last. As a result of the withdrawal of these trains, no passenger train was booked to use the GC connection onto the GN. Later that year, the famous double Whisker Hill Curve was rationalised to one track round the severe curve and crossovers into platform 1 and 2 were removed, the access round the curve now only from platform 3.

Sheffield Victoria closed to passengers on 5th January 1970, the passenger service from the Lincoln and Cleethorpes consisted of weekdays 0740 Harwich Parkeston Quay - Manchester Picaddilly and 1902 Cleethorpes - Sheffield Victoria via Retford and departures at 1544 (1437 Manchester Picaddilly) to Harwich Parkeston Quay and 0310 SuO (0117 Manchester Picaddilly) to Grimsby Town. The replacement service from Huddersfield to Sheffield ran to Nunnery Jn and reversed over the 1965 connection into the Midland Station.

The Harwich Boat Train was diverted away from its original route from 5th May 1973, and thus ended the Great Central and GN & GE Joint part with this famous train. An English Electric Type 3 hauled the "Boat Train" as it was locally known, into history. This also marked the end of the Sheffield to Gainsborough as major cross country route, now this line was relegated to locals to Retford, Lincoln and Cleethorpes, and every summer locomotive hauled trains returned to take holidaymakers to Skegness and Yarmouth.

During the early 1970s, local depots were reduced to one at Worksop, by the closure of Retford (Thrumpton) on 5th December 1970, followed by Woodhouse on 4th January 1971 and Darnall was the last to sucumb on 2nd December 1974.

Concentration of freight traffic over the Woodhead, saw upto 12 MGRs originating to the east and south, arriving at Rotherwood, and exchange of motive power to EM1 electrics. British Rail decided that Woodhead was obsolete and traffic could easily be transferred to other Trans–Pennine routes. Electric power was turned off for the last time on Saturday 18th July 1981, as a result Rotherwood Yard was closed. The up goods between Orgreave Colliery Signal Box and Woodhouse East Junction, and the down goods between Orgreave Colliery Signal Box and Darnall West Junction were taken out of use on 20th July. Orgreave Colliery ceased to produce coal in September, but B.S.C. coke to Scunthorpe continued. British Steels coking plant at Brookhouse (connection used old MR North Staveley branch) ceased coke production in October.The westend of Rotherwood Up

Yard was still used to shunt trains into and out of Orgreave Sidings until May 1982. The Coke traffic from Orgreave was transferred to British Rail through their ex Midland Railway connection. The loss of this freight closed Orgreave Sidings and enabled the remaining up and down goods loops to be taken out of use.

The introduction of HSTs in October 1982, onto the Midland line to London, caused a drastic reduction of empty stock workings into Darnall Carriage and Wagon Works, and other coaching stock continued to be maintained until May 1983.

Passenger services were withdrawn from Sheffield to Penistone via Nunnery Jn, from May 1983 and were diverted through Barnsley. The result of this was the line from Nunnery Junction through Sheffield Victoria to Sheffield Victoria No.3 signalbox was superfluous to requirements, and the track was removed.

The last colliery to close was Brookhouse on October 26th 1985, this left no colliery west of Kiveton Bridge, although traffic from North Derbyshire continues. The old junction at Nunnery was altered over two weekends in September 1986 (up line on 21st/22nd and the down line on 28th/29th), the line slewed onto the trackbed of the old mainline, and connected with the midland by MR milepost 159¾. This missed a bridge in poor state of repair.

Darnall Diesel maintenance depot was closed in October 1985, the work transferred to Tinsley. It was taken over by the wagon department, who had left the old steam depot. The carriage and wagon department was closed in October 1987 and transfered to Departmental use for maintainence of track machines. The old steam depot workshops were still being used for stripping of condemened multiple units.

Parkway Market saw a revival of traffic in the early 1980s with fruit from France arriving by rail. It arrived by trip freight from Tinsley Yard and had to reverse at Woodburn Junction, however by late 1987 traffic had disappeared and freight facilities were withdrawn from 8th February 1988.

Refurbishment of Worksop station started on November 1988, by stripping the slates and glass off the awning roof. All rotten wood replaced and new glass inserted. The present booking office to be moved to the opposite side of the entrance hall and set back, and approached through new entrance. Doors erected at either end of the entrance and converted into a waiting area. When not staffed this is locked. The awnings have been painted in white with the metalwork painted in dark blue with gold on the ornimented scrolls. The new booking office came into use on 19th June 1989.

The old steam depot workshop was taken over by Railease in October 1988, in order to repair their own wagons as well as other contract work. Darnall West Junction signalbox closed in February 1989, lately reduced to operating on one shift. The remaining up line from Attercliffe Junction becoming disused. Access to Darnall workshops now only available from Woodburn Junction.

Freight traffic is healthy from Brancliffe East Junction to West Burton Power station. Only Kiveton Park Colliery generates any traffic west of Brancliffe. Coal is still transported from Oxcroft and Markham and Renishaw Park (until March 1989). Between Woodhouse and Nunnery Mainline junction the only booked freight comes from Kirton Lime Sidings on Thursday and Saturday mornings and this traffic is threatened.

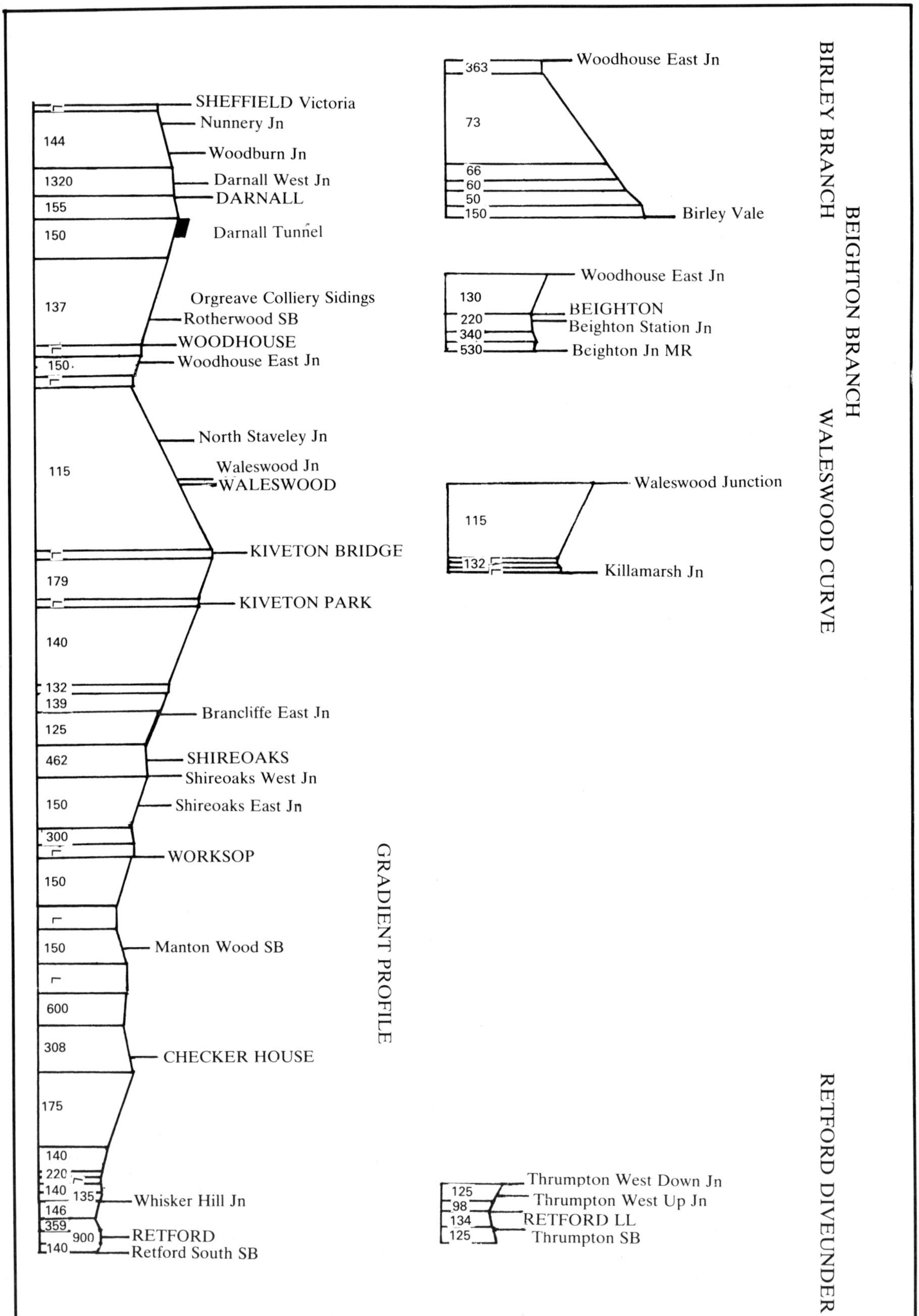

GRADIENT PROFILE
SHEFFIELD Victoria
Nunnery Jn
Woodburn Jn
Darnall West Jn
DARNALL
Darnall Tunnel
Orgreave Colliery Sidings
Rotherwood SB
WOODHOUSE
Woodhouse East Jn
North Staveley Jn
Waleswood Jn
WALESWOOD
KIVETON BRIDGE
KIVETON PARK
Brancliffe East Jn
SHIREOAKS
Shireoaks West Jn
Shireoaks East Jn
WORKSOP
Manton Wood SB
CHECKER HOUSE
Whisker Hill Jn
RETFORD
Retford South SB
144
1320
155
150
137
150
115
179
140
132
139
125
462
150
300
150
150
600
308
175
140
220
140
135
146
359
900
140
BIRLEY BRANCH
Woodhouse East Jn
Birley Vale
363
73
66
60
50
150
BEIGHTON BRANCH
Woodhouse East Jn
BEIGHTON
Beighton Station Jn
Beighton Jn MR
130
220
340
530
WALESWOOD CURVE
Waleswood Junction
Killamarsh Jn
115
132
RETFORD DIVEUNDER
Thrumpton West Down Jn
Thrumpton West Up Jn
RETFORD LL
Thrumpton SB
125
98
134
125

Sheffield Victoria sees B1 61211 on Platform 4 , sounding its whistle to depart on the 5.42 p.m. express to London Kings Cross on 4th July 1962. *(G.Coltas)*

GC C4 2915 (withdrawn June 1949) enters Sheffield with local train from Lincolnshire in the late 1940's. The two front coaches are articulated stock. *(H.C.Casserley)*

B17 61643 'CHAMPION LODGE' storms out of Sheffield Victoria with the Liverpool Central to Harwich Parkeston Quay boat train in 1955. Note the second coach is a Gresley buffet. *(R.Allsop Collection)*

Ivatt C1 ex LNER 2800 was withdrawn in July 1947, and restored externally to Great Northern condition and numbered 251. Steamed again in 1954, it approaches Sheffield Victoria with the 4.30 p.m. local from Lincoln Central on 18th August. This train was used as a running turn prior to working the Northern Rubber Company special from Retford to Liverpool Riverside on September 4th. *(D.M.Sutcliffe)*

The "North Country Continental" from Harwich Parkeston Quay to Liverpool Central approaches Sheffield Victoria hauled by Britannia 70009 'ALFRED THE GREAT'(30A) during September 1958, the first month of Britannia operation. The front two coaches complete with wooden boards, the Gresley buffet and front coach were removed at Sheffield, ready to return east with the Harwich train. *(K.R.Pirt)*

Blast Lane Goods Depot is host to a Modern Freight Exhibition during May 1958. Ex works BR 9F 92192 and English Electric Type 4 diesel D201 stand in the open air, and some stock is partially undercover. The passenger stock includes two Pullmans and two new maroon Mark 1's. *(K.R.Pirt)*

Brush experimental diesel D0280 'FALCON' passes Woodburn CEE workshops in March 1962 with the 1520 Sheffield Victoria to London Kings Cross 'Sheffield Pullman. Three new Met-Cam pullmans are included in the rake of coaches. *(Geoff Newall)*

B1 61193 (36A) hauls a six coach passenger train past Woodburn Junction Signal Box with a Hull express. The stock composed of three types of coach from two maroon Greslies and two BR Mark 1's one in maroon and the other in blood and custard livery. Note the brick base around the signalbox built in the Second World War to protect the equipment. *(K.R.Pirt)*

Class 40 080 heads 9 Mark 1's towards Darnall, with a Summer Saturday train from Manchester Picaddilly to Yarmouth on 6th August 1983. A brick wall was erected in the war to protect the base of Darnall West Junction Signal Box. In the background is Darnall Diesel Depot by then had become a Wagon repair shop. *(L.A.Nixon)*

Great Central up express headed by 8E 2–6–4 364 'LADY HENDERSON' passes through Darnall with a five coach train. A porter stands by two churns. The large water tank feed the water column on the end of the station. *(Sheffield City Libraries)*

(centre) Schools Class 30925 'CHELTENHAM' (70D) and 2P 40646 both whistle as they pass Darnall Station with a RCTS 'The East Midlander' 1X50 railtour from Nottingham Victoria to Darlington on 13th May 1962. *(T.J. Edgington)*

(below right) The westend of Darnall station as viewed from the Cresswell Road underpass approach on 25th May 1974. Although being unstaffed for five years, it managed to retain its champagne blue totems. *(M.J. King)*

(below left) Two ladies of Darnall station staff pose as though they were cleaning the enamel station sign and station seat. Note the neat flower beds that used to decorate this station. Today is just a sheet of tarmac with a bus shelter, minus fibreglass windows. *(Sheffield City Libraries)*

(above) The removal of Darnall Tunnel took place in 1912. This view shows the gradual stripping of the overlying sandstone at the westend of the tunnel. Steam cranes speeded up work. Rails laid on the ground to take the spoil away. *(Sheffield City Libraries)*

(above) Darnall Tunnel west entrance and Tunnel Signal Box viewed, before the removal of the roof and quadrupling. *(Sheffield City Libraries)*

(centre) Rotherwood Yard stabling point is a hive of activity on 1st November 1977, with the rear of the Beighton to Darnall C&W cripples passes on the down goods. In the down yard a pair of 76's is being backed onto a Lancashire bound MGR. While a single 76 and one 31 stable next to the offices. *(M.A. King)*

K3/2 61841 (fitted with GN tender) heads a 10 coach 'up' express through Rotherwood Yard. On the slow O4 hauls a freight. Two green EM1's are stabled in the headshunt. *(G. Coltas)*

The 1324 Skegness – Manchester Picaddilly Summer Saturday holiday train hauled by 40086, passes the sad remains of Rotherwood Yard on 14th July 1984. *(L.A.Nixon)*

Ian Allan 'Pennine Pullman' hauled by a pair of D11's 62662 'PRINCE OF WALES' and 62664 'PRINCESS MAY' approach Rotherwood Signal Box on 12th May 1956, where A4 60014 'SILVER LINK' will take over for the final leg to Kings Cross. *(D.M.Sutcliffe)*

Rotherwood Yard was used as an motive power changing point between the Woodhead electrics and steam, and later diesel locomotives. A O4/3 63889 has just passed Rotherwood Signal Box with an eastbound coke train. While B1 61186 stands just down the line from the water shoot, and O4/3 63664 await their next duty. *(P.Ransome–Wallis, National railway Museum)*

Just east of Rotherwood Yard K3 61867 (41A) approaches the A57 bridge with an 'up' express during September 1958. *(K.R.Pirt)*

Woodhouse station looking east from the down platform and shows the up station buildings, and were of a country station design. Composed of two waiting rooms, a sheltered area and a gents toilet. The footbridge only connected the two platforms. Gas lamps without station name light the platforms. *(Sheffield City Libraries)*

Britannia 70038 'ROBIN HOOD' (32A) sounds its whistle as it passes through Woodhouse station on 8th June 1960, with the 1.15 p.m Liverpool Central to Harwich Parkeston Quay 'Boat Train'. Parcels sit on the down platform awaiting the next train, while on the other platform posters advertise the delights of the seaside. The station manager was Mr K.H.Whitlow and could be contacted on Woodhouse 2436. (Geoff Newall)

"On 12th June 1965, an English Electric Type 3 D6745 (41A) is about to depart from Woodhouse with the 3.40 p.m. (2J05) Sheffield Victoria to Lincoln Central local, and a Derby Heavyweight unit (headcode 02) is about to pass Woodhouse West Junction Signalbox with a local from Retford. *(Geoff Newall)*

Cattle truck train leaves Woodhouse station past Woodhouse Sidings Signal box, on the left. Hauled by B16/2 61475 (30A) with Walschaerts valve gear during June 1958. *(K.R.Pirt)*

Triple headed Class 20's 64, 30 and 118 head the empty stock of the 'Three to the Sea' railtour between Woodhouse Junction Signal box and Woodhouse station on 2nd May 1987. MGR's from North Derbyshire to the Trent power stations, run round in the three sidings behind this train. *(L.A.Nixon)*

Beighton Station circa 1900 looking south towards the level crossing. The line to the left connects with the Midland Railway and straight on formed the London extension. Today only the signalbox (controls the level crossing barriers) and a modified connection with the 'Old Road' is all that remains. *(Heyday Publishing)*

(above) Waleswood Tunnel circa 1950 viewed from Killamarsh side. The distant operated from Waleswood Junction signalbox. The tunnel was obliterated in the extensive opencast mining in the 1970's. *(D. Ibbotson)*

(centre) Austerity 90250 (36A) approaches Killamarsh Junction round Waleswood Curve, with a long mixed goods from off the Worksop line on 16th April 1955. *(D. M. Sutcliffe)*

At Killamarsh Junction K3 61839 comes off the Waleswood Curve with a Grimsby to Swindon fish on 13th October 1956. *(D. M. Sutcliffe)*

English Electric Type 3 D6817 (41A) passes Brookhouse up distant with the 3.20 pm IB21 Sheffield Victoria to London King's Cross 'Sheffield Pullman' on 24th May 1963. *(Geoff Newall)*

Waleswood Station looking east towards Kiveton. This station was very basic, consisting of wooden platforms, a footbridge and two waiting shelters as well as a booking office. The colliery buildings loom over the station. *(D. Thompson)*

On July 16th 1907, Class 6B No. 424 and 2 coaches lie on there side down the banking. Passengers and crew survive. Superintendent Fowler of Gorton in charge of two steam cranes from Annesley and Gorton. Engine and stock removed on Sunday morning. *(G. Hurst Collection)*

The wooden platformed station at Kiveton Bridge sees the arrival of D11 62664 'PRINCESS MARY' on a local passenger. A man and a women step out to board the train. The station was light by gas into the early 1970's, and the platforms were replaced by concrete in the 1960's. After destaffing the wooden waiting rooms were replaced by open brick shelters. Its booking office survived for one year to serve the summer Saturday trains and then it was demolished.
(D. Thompson)

On 8th July 1979, Inter City 125 254020 passes Kiveton Park Colliery Signal Box with a diverted 0815 Newcastle – Kings Cross service. The brick base next to the box, signals is demise as it was a relay room for Kiveton Park Station box.
(L.A. Nixon)

Kiveton Park station as viewed from the westend, shows the main station buildings on the up line, containing the booking office and station masters house. The down platform two waiting rooms, sheltered area and a gents toilet. A footbridge connects the two platforms and survived until the mid 1970's. *(D. Thompson)*

Brancliffe East Junction Signalbox looking east towards Shireoaks on December 26th 1985. This box controls the junction onto the South Yorkshire Joint line. On the left a siding was provided for the Brancliffe Lime Quarry, which produced agricultural lime. *(G.Hurst)*

J11/2 5304 (later 4419) enters Shireoaks station with an up coal train in 1927. The gas light has the station name painted at the top. Shireoaks Lime quarry siding diverges to the left. *(P.Ransome Wallis, National Railway Museum)*

Looking east along the up platform of Shireoaks Station, shows that the station has altered little since being built. Gas lamps light the platform, the footbridge connects the two platforms in the middle of the station. A porter walks up the platform as the gates are wound open. In the background Shireoaks West box looms over the parapet of the colliery bridge. *(D.Thompson)*

Due to boiler problems at Cottam Power station, coal had to loaded straight into the furnace, at weekends and MGRs operated a seven day week. 58048 eases its 36 wagon MGR from Oxcroft on 11th June 1989, off the ex MR Mansfield branch onto the GC at Shireoaks East Junction. Normally these trains would run via Woodhouse, but the line was closed. The train was diverted via Elmton & Creswell and had to be banked to Clowne. Shireoaks East Junction signalbox still has a reasonable selection of semaphores, two signal gantries control the down lines. *(G.Hurst)*

Great Northern D3 1344 passes through Worksop Goods Yard with a Sheffield to Lincoln local in 1924. Note that behind the loco only fields exist. The siding to the left connected General Refractories and Shirley Aldreds to the mainline. *(P.Ransome Wallis, National Railway Museum)*

Worksop Sidings Signalbox closed at the end of the night shift on Saturday 7th December 1985. The previous day sees an 08 heading 2 Russel containers (to Scotland) and a VTG wagon (for Germany), and a 56 stands in the up sidings. Signalling for the down and up sidings was removed and replaced by stop and await instruction boards. Control passed to the Yard supervisor and movements radioed to either Worksop West or Shireoaks East Junction signalboxes. *(G.Hurst)*

Cravens 105 (LN) approaches Worksop station with a Saturdays Only 1456 from Sheffield to Retford in February 1984. In the week the train this terminated at Kiveton Bridge. Worksop West Signalbox is the last remaining MSL box in use. In the background MGR's extend round to the MGR maintenance depot. *(Geoff Hurst)*

31311 departs with 7 Mark 1's away from Worksop past its Goods depot with the 1826 Skegness – Sheffield. This locomotive hauled this train on 26th, 27th and 28th July 1983. The goods depot was later demolished during November/December 1985, and most of the bricks were exported to USA. *(G. Hurst)*

Bunting and flags festoon Worksop up platform, and two constables patrol the westend of the down platform, this meant the arrival of important dignitaries, a common occurrence at Worksop. As it formed the railhead for the estates of the Dukes of Portland, Newcastle, Lord Saville and the Marquis of Granby. *(Lens of Sutton)*

A Royal Hotel carriage pulled by two horses leaves Worksop station with a passenger. The station buffet was added in the early 1900's and blends with the 1848 buildings. A porch protects the carriages waiting next to the entrance hall. The Great Central Railway Goods and Grain Warehouse stands out in the background, in this 1909 postcard. *(G. Hurst Collection)*

Worksop Station during the early 1900's as viewed from the down eastside. In the background a Midland Railway train from Nottingham is seen arriving. The roof having been relined with glass gave a light and airy view. The large clock gives the time 7.30 and stood over the entrance hall, next to it is a W.H.Smith bookstall. The water column on the up side was used well into the 1960's. *(Lens of Sutton)*

A view taken from the old footbridge east sees three trains. The WD 90514 (36E) tender first hauls an up coal train. In the ABM sidings stands three vans and two coal wagons, while in the down siding bunker first Stainer 2-6-4T 42588 (16D) awaits its next duty on the Worksop to Nottingham service. Brush Type 2 D5599 (41A) is given the signal to enter with a westbound local to Sheffield Victoria, on 22nd August 1964. *(H. B. Priestley)*

Manton Wood Signal box looking from the northside of the line. This box disappeared in the late 1970's and control was transfeered to Worksop East. *(Chris Booth)*

GREAT NORTHERN

E 3201 (HD)

ILLUMINATIONS

EXCURSION

TO

BLACKPOOL

SATURDAY OCTOBER 21st

OUTWARD JOURNEY			Return fares second class	RETURN JOURNEY		
		a.m.	s. d.			p.m.
Gainsborough (Central)	dep.	10 0	20/3	Blackpool (Central)	dep.	9 55
Retford	"	10 21	18/9			a.m. (22nd)
Worksop	"	10 37	17/3	Sheffield (Victoria)	arr.	12 30
Kiveton Park	"	10 47	16/6	Darnall	"	12 42
Kiveton Bridge	"	10 52	16/-	Woodhouse	"	12 51
Woodhouse	"	11 3	15/-	Kiveton Bridge	"	1 3
Darnall	"	11 11	15/-	Kiveton Park	"	1 8
Sheffield (Victoria)	"	11 24	15/-	Worksop	"	1 18
		p.m.		Retford	"	1 37
Blackpool (Central)	arr.	2 20		Gainsborough (Central)	"	2 0

LIGHT REFRESHMENTS AT MODERATE CHARGES WILL BE AVAILABLE ON THE TRAIN

LATE BUSES ON RETURN AT :—

SHEFFIELD (bus tickets must be purchased with rail tickets)

Sheffield Transport Department buses meet the return excursion for the following districts: Abbey Lane, Arbourthorne, Attercliffe, Bents Green, Beauchief, Brightside, Broomhill, Cricket Inn Road, Crookes, Crosspool, Darnall, Ecclesall, Fulwood, Firth Park, Gleadless, Graves Park, Handsworth, Heeley, Hillsborough, Hollythorpe Rise, Hurlfield, Intake, Lodge Moor, Malin Bridge, Meadowhead, Middlewood, [illegible], Nether Edge, Nether Green, Parson Cross, Petre Street, [illegible], Prince of Wales Road, [illegible], Shiregreen, Sheffield Lane Top, Southey Green, [illegible], Wadsley Bridge, [illegible], Wincobank, Wisewood, Woodhouse, Woodseats, Woodthorpe and [illegible].

Fare :—Adults 1/3, Children **8d**.

PASSENGERS RETURN ON DAY OF ISSUE ONLY AS SHOWN ABOVE

Tickets can be obtained IN ADVANCE at stations and [illegible]

London, September 196[illegible]

Checker House Station at the turn of the century looking east towards Retford. Four members of staff pose for the photograph as well as the signalman in his box. One man carries a shunters pole and two others oil lamps. The station was very basic , but managed to have two waiting shelters. Today the A1 crosses over the line and only the platform railings survive through the undergrowth. *(George Matthews)*

Langwith Junction (40E) WD 90492 passes Rushey Siding Signal box on 14th September 1963 with a down coal empties. The stop block forms part of Rushey Public Delivery Siding. *(H.B.Priestley)*

31446 climbs away from Retford Low Level with the 1536 Lincoln Central to Sheffield local, up the bank to Thrumpton Junction on 20th September 1986. The line in the background is now single and was the old mainline, the down spur connects Retford High level with the diveunder. Note the sanddrag to retard any train that gets away. Normally this train was composed of 3 coaches. *(G. Hurst)*

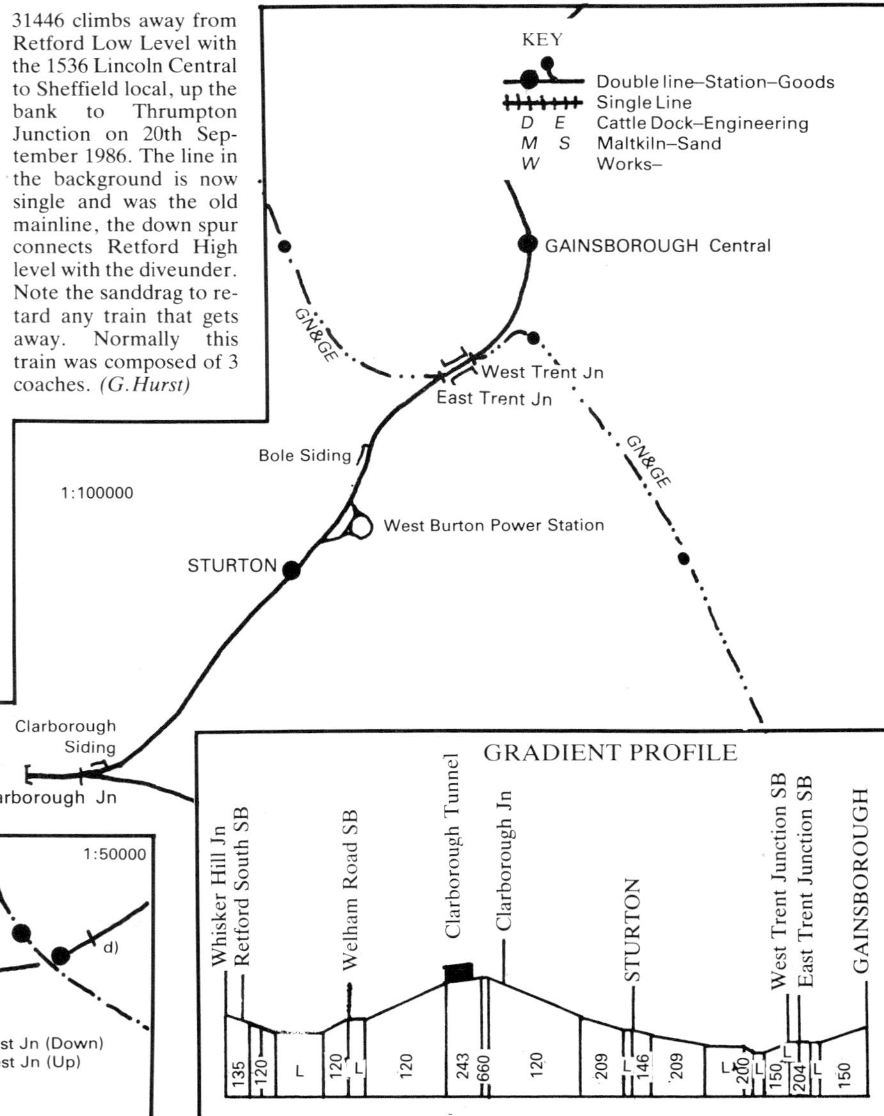

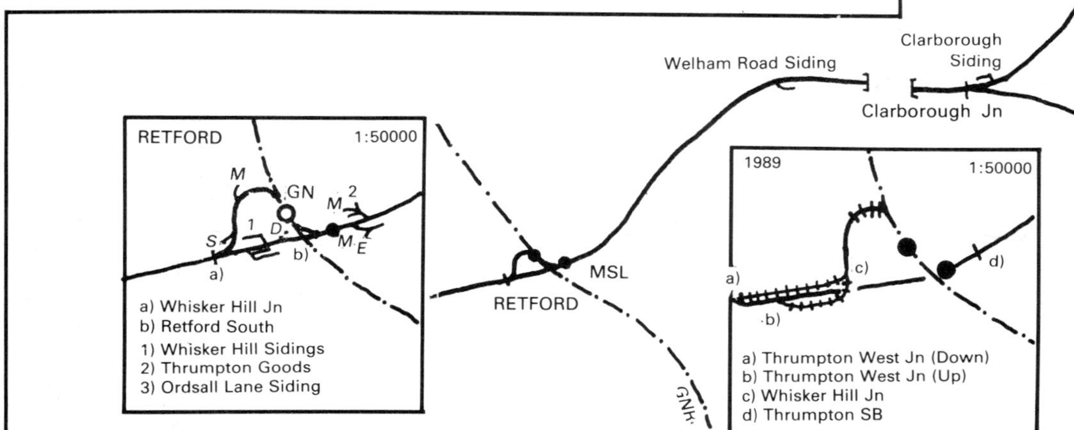

Whisker Hill Junction Signal Box is approached by a man with a bucket on his arm. A B1 with two vans and a brake come off the Whisker Hill curve. The direct line avoids the station, the down yard is packed with coal wagons and magnesia flasks, while the up yard is virtually empty on 19th July 1963. *(H.B.Priestley)*

Immingham Robinson B3 6167 (rebuilt in June 1938 with Caprotti Valve) eases away form platform 3 of Retford Station with a westbound local passenger train on 6th May 1935, composed of 4 coaches (of which 3 are celestory stock) and 5 fish vans. *(E.R.Morten)*

K3/2 61839 (40B) leaves Retford with a cloud of exhaust in 1953 on a Retford to Cleethorpes local. Platform 2 is occupied by an East Coast express and Platform 3 sees the tail end of a Sheffield train. Note the water column and cylinder to hold the shoot. *(R.Allsop Collection)*

Ordsall Crossing Signal Box on the original MSL route, now the Retford avoider on 19th July 1963, sees the approach of O4/1 63611 (41H) on an eastbound coaltrain. In the backgound the up sidings contains two lines of wagons, while the down yard has a brakevan and four wagons. *(H.B.Priestley)*

Retford South Signal Box is passed by an eastbound mineral train headed by WD 90220 (41F) in October 1958. The signal box was controlled by the Great Central who had right of way as the Great Northern arrived at a later date. *(K.R.Pirt)*

D11/1 62662 'PRINCE OF WALES' (41A) eases out of Retford, past a fine signal gantry onto the Sheffield to Grimsby mainline with the 11.30 a.m. to Cleethorpes on 1st August 1959. *(R.J.Buckley)*

56028 'West Burton Power Station' traverses over Retford diveunder and is about to pass Thrumpton box with a MGR to Cottam Power Station, on a very hazy evening in May 1989. Thrumpton box was converted from manual to colour lights and replaced 5 signal boxes in June 1965. *(G.Hurst)*

The former MSL passenger sta tion clock shows 2.55 p.m and i passed by Class 2 4–4–0 700 on 1.00 p.m Cleethorpes to She field slow. An unidentified 9E 0 6–0 shunts an express goods nex to the Retford MSL shed durin 1899. *(Pegler Collection, Basse law Museum)*

Welham Road Signal Box is about to be passed by English Electric Type 3 D6751 and Brush Type 2 D5691 both 41A on an empty coke from Frodingham to Brookhouse or Orgreave on 30th March 1964. *(H.B.Priestley)*

Early 1900's view of Sturton Station, with two ladies waiting for a Retford train. The main building is of a cross shape, of which all third class stations were originally designed. In the background, the goods yard is well filled, with coal wagons on the coal hopper. The local pickup goods stands on the up waiting the crossover onto the main down. *(G.Hurst Collection)*

et-Cam Class 101 (rear carıge ex Scotrail) passes ainsborough Trent Junction gnal on 1st May 1989, with the .27 Sheffield to Cleethorpes cal. The brick signalbox in the ıckground was built in 1960 reacing the second box on the ɪrthside next to the bridge. *ȷ.Hurst)*

East Trent Junction Signalbox straddled the line on an iron trestle bridge (West box constructed in same manner). An accident demolished the support piers and a brakedown crane is helping in the rcmoval of debris. (Jim Jackson Collection)

2. Gainsborough – Wrawby Jn

The Great Grimsby & Sheffield Junction Railway (GG&SJ) submitted plans for its mainline from Bole to Grimsby in November 1844. The line formed an endon junction with the Sheffield and Lincolnshire Junction railway, thus completing a railway line from industrial Sheffield to the port of Grimsby.

The railway from Bole to Barnetby was authorised in the Great Grimsby and Sheffield Junction Railway Act of 30th June 1845, :-"Making a railway from Parish of Bole in the County of Nottingham, near Town and Port of Gainsborough to the Town and Port of Grimsby.

Junction with intended Sheffield and Lincolnshire Junction Railway.

Crossing of turnpikes on the level in parishes of Scawby cum Sturton Wrawby 2; Blyton cum Wharton 1; Northorpe 1; Hibaldstowe 2; and Bigby 3.

Provide against obstruction of River Trent. One opening of at least including holing path of 100 feet, 10 feet clear of part holing path.Ancholme Navigation height and width of bridges set in act.

Period of completion 7 years. The line to be double to Brigg, but single to Gainsborough. But must be double within 12 months of opening to public".

The board of GG&SJ awarded the contract from Bole to Grimsby to John Stephenson & Co in Autumn 1845 for an estimated £389,929 and to be completed by 1st May 1848.

The '*Stamford Mercury*' on 3rd July 1846 gave the news :- "That excavations of 40 feet deep had been made for Trent Bridge foundations".

The Manchester, Sheffield and Lincolnshire Railway (MSL) board were told on 10th February 1847, that the heavy earthworks at Gainsborough and Blyton more than removed. The bridges and culverts on this line are all built. Two extra shafts have been sunk for Kirton Tunnel, and about a quarter of the tunnel bricked and finished. Two large bridges over the Ancholme Navigation are progressing rapidly.

The '*Doncaster Gazette*' of 9th April 1847 reported :- "Great number of piles driven at Gainsborough for Trent bridge, but orders were received that work should be suspended. The bridge according to present plans was to cross the river at an inclination of 40 degrees. The Trent Navigation company have announced it should cross at right angles. To effect this it is proposed to divert the course of the current into a new channel and cut off the long curve in the river. The matter is under deliberation. The plan from Mr Lister of the contractors was to cross 100 yards further down and this would enable to cross at 45 degrees". It was not for six weeks that the same paper reported that :- Works on the railway over River Trent have resumed after interruption caused by Tidal Commissioners objecting to crossing of Trent by bridge as interfering with navigation. The alteration to direction is trifling, instead of crossing at 40 degrees but 50 degrees and supported by girders and three piers at a distance of 150 feet from each other. This plan of crossing by girders and the piers being sunk like the bridge over the Dee at Chester, no anxiety is felt in Gainsborough over safety".

Work progressed on building the line and in July 1847 '*Stamford Mercury*' reported, on 3rd :- "Men now employed on construction of bridge over Lea road and a short distance from the River Trent", followed on 24th :- "Railway made through playground belonging to Crammers School at Gainsborough, and its area has been reduced in area by one half, cutting the view with a bank of 16 feet high". The next news came on 31st with :- "Bridge over Lea road beautifully faced with with a stock brick and is to be an iron bridge. Work delayed near grammar school due to tardiness with purchasing land adjoining. May be also a delay due to shortage of rails".

A later report to the board on 18th August, stated that excavation from Gainsborough to Kirton Tunnel will be finished next month and that 800 yards of tunnel bricked and completed, with 500 yards remaining.

The local press reported the progress and difficulties of the construction of the line. The mining of Kirton tunnel was finished Tuesday last (13th March) with exception of a cutting a little beyond Kirton, the heavy work on the Gainsborough line between Kirton and Gainsborough is finished. A fortnight later the last brick was laid in Kirton Tunnel in the presence of Mr Fowler as well as contractors Messrs Stephenson who the press reported :- "Have spared no expense in completing the work in proper style. At the finish Mr Fowler surveyed it with strictest scrutiny and publicly declared it finished and gave immediate orders that 130 workmen at his expense have 1 gallon of ale each.Messrs Stephenson added that a first rate supper for every individual belonging to the work". The '*Stamford Mercury*' reported on 26th May, "That after great delay, the works of railway station at Gainsborough are commenced by felling trees, clearing away the ground and collecting materials. Considerable interest what kind of station the united railway companies are going to provide. Six weeks later the paper said that Messrs. Kirk and Parry contractors for the stations are pushing forward with the works on the top of Spring Gardens whilst excavations, accommodation work etc. at Pringle Hill, remain a perplexing puzzle to everyone. In just over two months, the passenger station top at Spring Gardens is now covered in and the goods station extending towards Spital Terraces will be ready for its roof in a few days. Preparations are being made for coal and lime depot. Nothing yet transpired on opening of line to Grimsby. Hoped that it would be available prior to Prince Alberts visit to Grimsby".

The '*Stamford Mercury*' of 29th September 1848 recorded that :-"On Friday last (22nd September) an engine and two first class carriages arrived at the new station at this place (Brigg), which is with large and spacious wharehouse are fast progressing towards completion. It having been known on the previous evening that an express train would be at Brigg at 8 am on Friday for purpose of making trip to New Holland, crowds of gratified persons rallied to the station point and were gratified at seeing special train conveying Earl and Countess of Yarborough, M.B.Luard Esq., Mrs Luard, G.M.William Esq. of Brocklesby, Mr Stephenson the contractors, Mr Potter the engineer. The train stopped at Brocklesby etc. the line will be shortly opened for general conveyance".

Ulceby to Brigg opened on 1st November 1848, and the first timetable is given below in Table 1 :-

Weekdays							Sunday			
	am	am	pm	pm	pm	pm	am	am	pm	pm
HULL		815	1215	315	515		815		515	
NEW HOLLAND		9 0	1245	345	445		845		545	
MARKET RASEN	743		1243	343		634		1058		643
BARNETBY c										
BRIGG	838	950	135	435	635	728	935	1148	635	728

Weekdays							Sunday			
	am	am	pm	pm	pm	pm	am	am	pm	pm
BRIGG	8 0	920	1 5	4 5	6 5	7 0	9 5	1115	6 5	7 0
BARNETBY c										
MARKET RASEN		1015	2 0	5 0	7 0		10 0		7 0	
NEW HOLLAND	856		156	556		756		1211		756
HULL	926		226	626		826		1241		826

c change at Barnetby to or from Brigg

One week after the opening of the Brigg to Ulceby line the Eastern Executive Committee inspected the works, over a period of three days, and noted that "Gainsborough station commodious, and Brigg station road unsatisfactory".

The '*Stamford Mercury*' of 16th February 1849 stated :-"The station at Gainsborough is nearly finished except ballasting and laying permanent rails. It also thought that the line would not open until 2nd April, but it was reported that the painters were working on a Sunday in order to finish the contract".

At the end of March, Gainsborough station was crowded, in anticipation of the arrival of a loco. It did not arrive till Wednesday about noon. The well known whistle was heard at some distance, and by hundreds of persons who hastened to the station. A veritable loco on a line of railway with a long train of wagons laden with cliff stones for ballast. Considerable apprehension had prevailed respecting safety of bridges, and some alarmists went so far as to say that they had given way. We are authorised to state that they stood well, these suffered a deflection of ½ an inch and are regarded as trifling in such girders. Since 21st engines have been employed working night and day. On Saturday night the permanent rails, a double were about completed.

The 'Sheffield Independent' stated that the line between Gainsborough and Brigg had been inspected on 29th March 1849 and been found in a very satisfactory state. It opened to the public on 2nd April and the '*Doncaster Gazette*' reported :- "On Monday, the railway between Gainsborough and Brigg opened for traffic, having been inspected by Captain Wynn on Thursday 29th March and found it in a very satisfactory state.The day was observed as a general holiday, all business being suspended. It is hoped that Lord Yarborough would accept an invitation to *de jeuner* which was sent to him and the other directors, and also the engineers and contractors. His lordship being in London making arrangements with Prince Albert relative to laying the foundation stone of the Grimsby Dock, was unable to attend. Four directors, Messrs. Heneage, Blake, Williams, and Dr. De Bartolomi, arrived by a special train at two o'clock, and were met at the station by the principle inhabitants of the town (Gainsborough). They went by procession to the White Hart Hotel, where a superb luncheon had been provided by Mr and Mrs Parker".

A Stamford Mercury reporter was allowed to travel over the line :- "The line maintains the reputation of Mr Fowler, the engineer, in planning and Messrs Stephenson in making the line. The cuttings at Gainsborough and Blyton and the tunnel at Kirton are the heaviest portions of the works; the arch of the tunnel and the bridges are firm and well finished. Monday was quite a gala day in Gainsborough: all the shops were closed, and the day was observed as a holiday. The weather was propitious till evening, and hundreds of persons crowded the station".

The timetable and some fares are given in Table 2, given below :-

Weekdays					Sunday		Fares		
	am	am	pm	pm	am	pm	1st	2nd	3rd
HULL	6 0	915	230	545	815	545			
NEW HOLLAND	620	945	3 0	615	845	615			
GREAT GRIMSBY *	*459*	*845*		*530*	*845*	*532*			
BARNETBY	648	1015	330	645	930	645	*2/7½*	*1/11*	*1/4*
BRIGG	7 0	1027	342	7 1	942	7 1	*3/3½*	*2/5½*	*1/8*
SCAWBY & HIBALDSTOW	7 8	1035	350	7 9	950	7 9	*3/9*	*2/9*	*1/10½*
KIRTON LINDSEY	717	1044	359	718	959	718	*4/4*	*3/2½*	*2/2*
NORTHORPE	724	1051	4 6	725	10 6	725	*4/9½*	*3/7*	*2/4½*
BLYTON	732	1059	414	733	1014	733	*5/3½*	*3/11½*	*2/7½*
GAINSBOROUGH	742	11 9	424	741	1024	741	*6/2*	*4/7½*	*3/0½*

* Change at Ulceby

Weekdays					Sunday		Fares		
	am	pm	pm	pm	am	pm	1st	2nd	3rd
GAINSBOROUGH	740	1230	225	551	1040	551			
BLYTON	751	1241	236	6 2	1051	6 2	*0/9½*	*0/7½*	*0/4½*
NORTHORPE	759	1249	244	6 8	1059	6 8	*1/4*	*1/0½*	*0/10*
KIRTON LINDSEY	8 6	1256	251	615	11 6	615	*1/9*	*1/4*	*1/1½*
SCAWBY & HIBALDSTOW	815	1 5	3 0	624	1115	624	*2/4½*	*1/9½*	*1/1½*
BRIGG	823	113	3 8	632	1123	632	*2/10½*	*2/2½*	*1/4½*
BARNETBY	834	124	319	649	1134	649			
GREAT GRIMSBY *	*929*	*210*	*442*	*812*	*1221*	*736*	*6/0*	*4/6½*	*2/10½*
NEW HOLLAND	9 7	157	352	722	12 7	722	*5/8*	*4/3½*	*2/8½*
HULL	937	227	442	752	1243	752	*6/2*	*4/7½*	*3/0*

* Change at Ulceby

The junction of the MSL and GN lines at Gainsborough was inspected by Captain Wynne. It was intended to open the GN line from Lincoln on 2nd April. However Captain Wynne was unable to write his report, as he had to go to Manchester to investigate the London North Western Railways blockading of the MSL station. This meant that the opening had to be deferred until 9th April. The GN did not have its own station at Gainsborough, but used the MSL station. This involved reversal at Trent Junction. Four trains operated each weekday and two on Sunday. Connections were poor at Gainsborough, and caused the residents of Kirton to complain that the first train from Kirton, missed the Lincoln train by 15 minutes.

One month after the opening of the line to Brigg the Lincolnshire Chronicle reported that :- "Preparations are being made for a coal and lime station at Gainsborough, the road will be from Beaumount Street, off Sandfield Lane. There will not be a supply of articles until the line from Sheffield is opened in July next".

The Eastern Executive were informed on 20th February 1850 of the final building costs for the stations and other works. Gainsborough Station and Fittings £6,806 0s 0d, platform and iron roof £4,107 0s 0d, Carriage Shed £510 0s 11d , Goods Station £5,369 0s 4d and Engine House £2006 12s 2d; Kirton Lindsey Station £1,588 16s 8d and Goods Station £2,203 12s 6d; Scawby & Hibaldstow £1,326 4s 5d; Brigg Station £4,468 14s 11d and Goods Station £2,788 14s 2d. Blyton and Northorpe were classed are 3rd class stations and cost £420 each plus £280 18s 8d for platforms and fences.

Cast iron bridges at Blyton and Thonock Lane were declared as unfit, in 1858. The company ordered on 20th January, that new girders be cast at once and a report on the cost be made.

Marshalls were the only firm in Gainsborough to have private sidings. As this firm grew, so did the sidings required. The company first asked for two sidings on 20th October 1871. The board estimated these would cost £500, and were approved at a meeting on 17th December. Two years later on 11th October, the company asked for there sidings to be extended. The MSL agreed to the request and estimated the cost to be £110.On 15th September 1876, the MSL agreed that Mr Gambles tenancy of Gainsborough refreshment rooms should continue.

In the 1880's, the MSL traffic grew and extra sidings had to be provided. The sidings at Kirton Lindsey were to be rearranged at a cost of £300, the board approved the expenditure on 23rd April 1880. For £140, on 17th November 1882 the MSL decided to build goods offices for Gainsborough. Less than two weeks later, Mr Matthew Pitts won the tender to renew a timber bridge for £2,633. Scawby & Hibaldstow goods traffic had grown sufficiently that it required a goods shed, and the board approved its construction on 23rd February 1883 for £65. Gainsborough hosted the Lincolnshire Show that year and the MSL agreed to provide the necessary accommodation for the expected surge of traffic, and it was estimated to cost £550. On 15th January 1886, the company agreed that work costing no more than £1,900 was to be allowed for the building of a subway for Sandsfield crossing, Gainsborough. Eleven months later the plan was altered into a bridge with Mr Cowan Smith winning the contract with his tender of £1,486. Marshalls approached the MSL with a view to building a 270 yard siding, next to up main line. It was estimated that this would cost £336, with Marshalls paying £200 of the costs. The erection of a footbridge for Gainsborough station was sanctioned by the company on 9th November 1888 for an estimated sum of £300. Traffic continued to expand at Scawby, that on 30th January 1889, the board approved an extension to the wharf and goods sidings. The last major work of the decade occurred at Northorpe with additional sidings approved on 20th December for an estimated £1,125.

The next ten years the improvements and extensions continued. A footbridge for Brigg Station was approved on 13th March 1891, it was estimated to cost £350, and at the same time additional sidings accommodation was approved to cost £538. During this decade saw a great change at Gainsborough, the Cattle Market was expanding and the MSL sought to increase its traffic, the board were told on 26th February 1892 that new pens and an extension of the siding were required at approximately £1,219. Three months later, the Cattle Market reported to the MSL board that the proposed alterations were no good. They proposed to build the cattle pens themselves for a nominal rent. Brigg stationmasters house was located at the eastend of the station, however the company approved expenditure of £214 on 6th January 1893, to alter the westend house to become stationmasters accommodation, the east house to be changed to accommodation for staff. Work was approved by the board to rearrange the siding at Gainsborough to allow the building of a cartway, and this was estimated to cost £540.

Part the MSL Act of 6th July 1895, authorised the following work :-

A railway No.10 18.8 chains in length being a widening of Retford to Gainsborough over River Trent commencing by a junction with that railway in Parish of Bole and terminating by junction with same railway in Parish of Gainsborough.

Gleadalls were to build a maltkiln at Kirton Lindsey, and asked the company to provide a siding, the board met on 27th January 1897 and approved this request. It would cost £510 and Gleadalls would have to pay £300 towards the cost. Later that year, the local clergy asked the MSL to provide a footbridge at the station, the board agreed to postpone a decision. Suddenly the board on 12th January 1898, approved the erection of a footbridge at £345, which was now urgently needed. Mr Jeffery of Retford opened a malt kiln at Blyton, and on 2nd June 1899, the company authorised the expenditure of £347 10s for a crossover. Messrs Parrys Lime Works sidings at Kirton, were altered for more accommodation for shunting wagons, the expenditure of £119 was authorised by the company on 27th October 1899.

The '*Lindsey & Lincolnshire Star*' of 20th April 1901 reported that "Owing to the great increase in traffic, the Great Central Railway have decided to widen and strengthen the bridge over the Trent at Gainsborough. The existing structure consists of two spans of 170 feet and two side spans, or flood arches. The bridge will be widened on the southside, only necessitating the enlargement of the foundations, piers, abutments, etc. The strengthening is to be carried out by inserting a central girder, to which the existing girders will be bolted. The contract has been let to Cleveland Engineering Company, Darlington, who have already started operations, under the supervision of Mr J.F.Coltas the district engineer of the Great Central Railway".

In 1901 the Lincolnshire Show was held at Brigg, and the company provided the additional facilities to cope. Additional siding £875, £290 for filling, £325 for permanent way, £50 for maintenance, £135 for a wharf, £11 for a gate, Carriage Dock £2 and £62 for additional signalling.

A banking engine was kept at Kirton Lindsey and the GCR board approved the building of a lay bye siding for £356, on 20th February 1903. That same year Marshalls of Gainsborough, complained that its sidings were frequently blocked in daytime. The company estimated that additional sidings would cost £4,135. They also found out that loading in darkness would cost £250 (including a 10 ton crane), plus £15 for electricity. Marshalls provide carriage of £12,000 per year.

It was not until 20th May 1904, that the company decided to build a subway from the station and a verandah, to enable an extension to Marshalls siding. T.E.Johnsons tender of £2,006 16s 9d was accepted by the board.

From 1904 to 1910, the GCR had to undertake repairs to bridges, the first bridge was at Brigg which needed strengthening, and only three years later the River Ancholme bridge at Brigg, required a major overall. The last work was undertaken at Northorpe, where Rolletts bridge was renewed for £130 13s.

The outbreak of the First World War brought little change to the fabric of the railway. In 1918, the War Office built a filling factory near Thonock level crossing. The War Office was to pay 50% of the cost of installment of sidings. The board minuted this request on 9th January, and were informed on 9th April, 'that whole of permanent way laid, except extension of up and down layby siding. The work was completed by 10th June, however due to increase of traffic over level crossing, additional siding and new gates necessary. The work was under consideration.

The overall roof of Gainsborough station was removed and replaced by to separate awnings. The decision to rebuild was taken at a Great Central Traffic Committee meeting on 21st January 1921. The contract went to Mr Thomas Carr on 2nd March, the committee minuted that "contract documents, not yet signed, owing to certain preliminaries having to be arranged, but do not anticipate any difficulties". On 24th May, the material for Gainsborough, was in contractors yards and a portion delivered to site. By 21st June, over 50% of old roof striped and one month later, the roof was totally striped. Foundations for new roof complete and steelwork was all at the site. The new roof was reported as complete at the board meeting of 21st November. This being the last works to be completed by the Great Central Railway on this line.

The Great Central Railway was grouped to form part of the London and North Eastern Railway from 1923. The new traffic committee met on 5th July 1923 and recommended that Gainsborough North Signal Box should be closed and its work taken over by Gainsborough South. The company would save £111 annually, but the initially would cost £900. Gainsborough Central station platform at 300 feet long was not long enough, and at the Traffic committee meeting on 3rd January 1924, it authorised an extension of 150 feet to cost £1,025. The water column to be taken down and rebuilt in a new position. The LNER deposited plans in November 1924 for a new railway from Northorpe to Frodingham, and passing through the following parishes :-Scotton; Scotter; Messingham; Bottesford to Frodingham.

The LNER Act of 31st July 1925 authorised construction of :-

"Railway No.3. 9 miles 36 chains, commencing parish of Northorpe, junction with Sheffield and Grimsby Railway near Northorpe station and terminate in Urban District of Scunthorpe and Frodingham by junction with Barnsley and Barnetby railway, 13 chains west of bridge carrying Ashby Road over said railway.(See Chapter 2, Volume 2 for more details.) This railway was never built.

Brigg Sugar Factory was the last siding to be connected to the main line. It was built in 1928 and came into production with the 1929-30 season.

The Locomotive Committee requested the Works Committee to augment the water supplies at Kirton Lindsey with a softening plant. The Works committee estimated it would cost £6,849 and ordered on 7th January 1937 that tenders be obtained. It was over a year on 17th February 1938, that the Works committee accepted the tender of £2,993 10s 10d from Fletcher & Co. for the provision of a reservoir, sludge pit, water mains and chemical mixing house, as a subsitute for water softening plant.

After nationalisation, the railway scene did not change a great deal until the the mid 1950s. Northorpe was the first station to lose its passenger services on 4th July 1955, but retained its freight for another 9 years. Blyton station lost its passenger service from 2nd February 1959. Motive power along this line changed firstly on passenger services with the introduction of diesel multiple units in the 1950s, the fast passengers to Manchester and Liverpool remained steam hauled into the early sixties, then being replaced by Brush Type 2 and English

Electric Type 3 diesels. Freight gradually pasted into the hands of WDs, before being totally dieselised in 1965. Freight facilities were withdrawn from all but Brigg and Gainsborough Central in 1964, Blyton and Northorpe totally closed to traffic on 2nd March, and then followed by Scawby and Hibaldstow from 4th May. Kirton Lindsey goods depot closed from 14th June that year.

The Sugar Beet season of 1967–8 produced special freights of Beet to Colwick 0945 MX and Dried Beet Pulp to Kidderminster 1105 WO. The next season saw traffic to Aberdeen and Montrose 1130 MO, Leith, Hawick and Inverkeithing WO, Aberdeen and Inverness FO.

Scawby & Hibaldstow station closed to passengers on 5th February 1968, its last train ran on the 3rd. Its final service consisted of 3 trains to Sheffield and 4 to Cleethorpes, each weekday. This timetable marked the end of this route being the main line from Sheffield to Cleethorpes, 7 trains each weekday operated from Sheffield to Cleethorpes each weekday until 4th May 1968, and from that day the service was reduced 5 trains each way, but kept the 3 trains on Sunday.

Freight trains from the Retford line to Frodingham had to reverse at Barnetby, with the engine running round or another engine forward. This resulted in delays and extra costs. The Eastern Region put forward a plan in 1964, to slew the mainline to the south and connect with the Lincoln line south of Wrawby Junction. Install a direct line, and creating a Barnetby avoider. This would enable the freight to run directly to Frodingham and reduce the facilities at Barnetby. The project did not reach fruition.

Brigg and Kirton Lindsey stations became unstaffed halts with the introduction of paytrains, however Gainsborough Central remained staffed until the mid 1970s. All local tickets were issused by the train guard, and initially the single fare was half of the day return. The after Gainsborough Central became unstaffed, vandalism and decay took there toll on the buildings. Deteriorated so far that they were demolished during the summer of 1977, and were replaced by two bus shelters.

The newspaper train from Manchester Piccadilly to Cleethorpes passed over the line daily was diverted to Lincoln from 2nd May 1977, and reduced the passenger service to 3 eastbound and 4 westbound and no regular Sunday service. The following year a summer Sunday train was restored for 3 months, but the weekday trains were reduced to 3 westbound trains. The state of the track had deteriorated and the line was infested with p.w slacks. slacks.

The 1980s brought great change to this line. Over a two year period the line was converted to three stretches of single line, one crossing loop at Northorpe and a long double line from Kirton Lime Sidings to Brigg Signal Box. From 17th May 1982, Northorpe SB to Kirton Lime Sidings was reduced to single. The up line was removed, however the track was slewed into the middle to pass under an over bridge west of Kirton Lindsey and through Kirton Tunnel. At Kirton Lime Sidings the track crosses over onto the up and then returns back to double east of the signal box. As a result of singling Kirton Lindsey up platform and footbridge became redundant and were quickly demolished. Stage two converted the line from 76 Miles and 6 chains (near site of Thonock Signal Box) to Northorpe Signal Box, using the down line. Northorpe became a bidirectional crossing loop, 50 chains in length and located east of the level crossing.The last section to be singled was from Brigg Signal Box to 62 chains west of Wrawby Junction Signal Box, on 18th November 1984. As part of the scheme, colour signals and electric token came into force.

The sole Sunday train ran for the last time in Summer 1986, starting on 20th July for six weeks until 31st August. It departed 0935 from Sheffield to Cleethorpes and returned at 1805. The train stopped to pick up only at Gainsborough Central, Kirton Lindsey and Brigg on the outward journey and set down only on its return.

Freight disappeared from this line during the miners industrial action in 1984–5. The night shift was taken off in February 1985. Coal returned, but not in the same quantities. In its heyday 6 MGRs ran, 3 from Worksop and 3 from Doncaster. Failure of the conveyor belt at Immingham, reduced the booked freight down to a MO Immingham – Doncaster MGR and ThSO Kirton Lime Works – Earles cement. Plus additional oil trains to Cottam Power Station and Tuxford to Immingham which used to run via Darnall West Junction and since its partial closure the occasional freight runs to Worksop Yard and 0700 Immmingham – Tuxford and 0700 Tuxford – Immingham swap engines. It usually runs on a Wednesday. If there are any problems between Barnetby, Scunthorpe and Doncaster passenger and freight is diverted over this route. ıte.

In November 1988, British Rail announced that it intended to withdraw passenger services over this line, and put the line into mothballs to await the projected coal imports form Immingham to the Trentside Power Stations. The closure notice appeared in March 1989, the poster stated that closure would occur on 4th October 1989, if no objections were received. It is proposed to replace the poor train service with a bus service from Gainsborough Lea Road to Barnetby running in the same times connecting with trains at either end. Objections have been received and the public inquiry was to take place in July, however the week before the inquiry, British Rail announced that it wished to close Gainsborough to Doncaster. This affected the objections and the inquiries were cancelled. Since mid October PowerGen have signed a contract to import foreign coal via Immingham and 5 MGRs are booked per weekday to traverse this line.

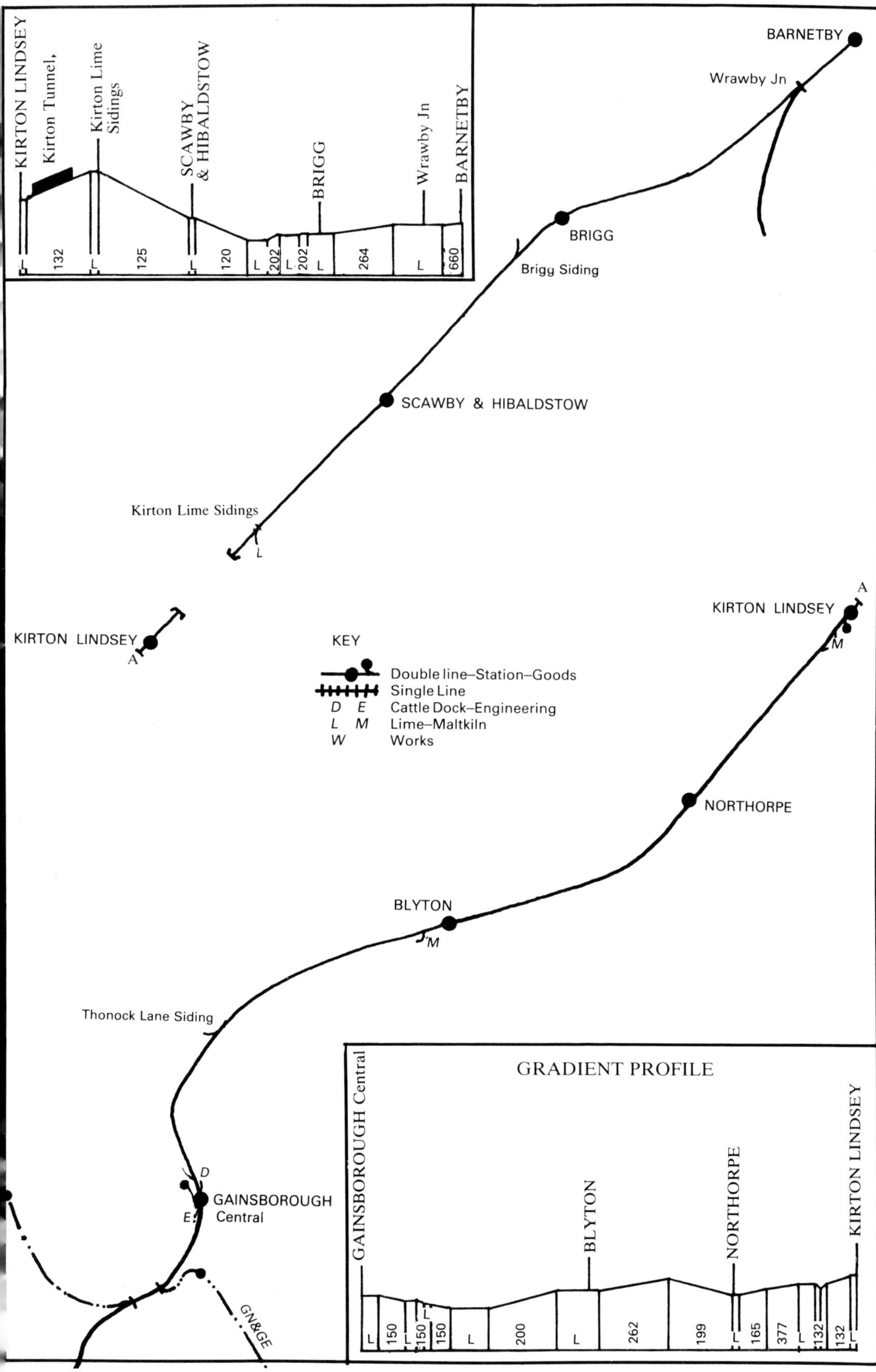
KIRTON LINDSEY
Kirton Tunnel,
Kirton Lime Sidings
SCAWBY & HIBALDSTOW
BRIGG
Wrawby Jn
BARNETBY
L 132 L 125 L 120 L 202 L 202 L 264 L 660
BARNETBY
Wrawby Jn
BRIGG
Brigg Siding
SCAWBY & HIBALDSTOW
Kirton Lime Sidings
L
KIRTON LINDSEY
A
KEY
Double line–Station–Goods
Single Line
D E Cattle Dock–Engineering
L M Lime–Maltkiln
W Works
A
KIRTON LINDSEY
M
NORTHORPE
BLYTON
M
Thonock Lane Siding
D
GAINSBOROUGH Central
E
GN&GE
GRADIENT PROFILE
GAINSBOROUGH Central
BLYTON
NORTHORPE
KIRTON LINDSEY
L 150 L 150 L 150 L 200 L 262 199 L 165 377 L 132 132 L

On 9th June 1947, a J11 4287 approaches Gainsborough Central station with a Retford to Cleethorpes local service. The stock is composed of Gresley coaches and a Southern Railway van. The sidings on are well loaded with of vans and wagons. *(R.J. Buckley)*

Kingsway postcard of Gainsborough Great Central station at the turn of the century. The roof covers over the platforms, glass in the roof allows a reasonable amount of light through. Beneath it a large number of passengers stand on the Grimsby platform waiting for the train to arrive from Retford. W.H.Smith have a bookstall on the platform, full of newspapers. Gas lamps hang from the roof. A locomotive stands on the *up* taking water from the column located at the end of the platform. *(Lens of Sutton)*

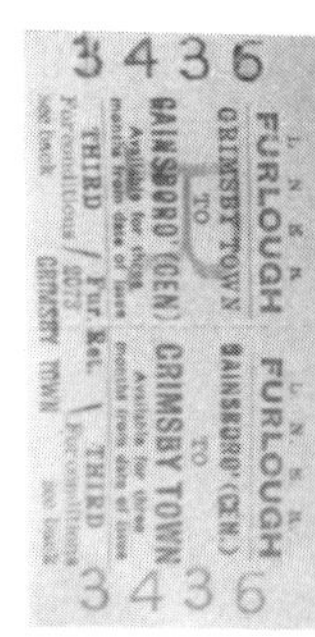

Gainsborough Central station front was built with these fine columns, however to improve the goods facilities it was necessary to construct a siding across the front. A subway was built to the approach road. *(H.B. Priestley)*

B1 1152 (with immaculate paintwork) stands in Gainsborough Central on 9th June 1947, with the 2.20 p.m. local from Cleethorpes to Retford, composed of six coaches of various design. *(R.J.Buckley)*

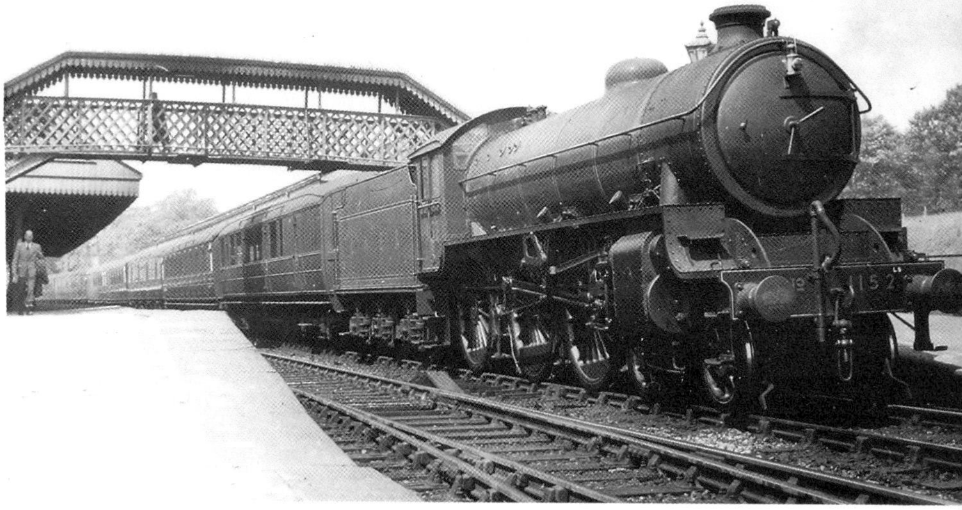

A van stands on the platform of Gainsborough Central station on 1st September 1970. Three barrows litter the platform. Behind the second barrow, the bow window was the site of the refreshment room. Three enamel signs survive locating the Stationmasters Office, Parcels and Left Luggage, and the Way Out. *(M.A.King)*

J11 5324 shunts its wagons and vans in Gainsborough Goods Yard on 9th May 1933. The building in the background is the Goods Depot. *(H.B.Priestley)*

Lincoln based Craven 105 (rear car E51298) stands in Gainsborough Central with the 1838 Sheffield to Cleethorpes on 10th July 1986. Only the footbridge survives of the original station. Two bus shelters are the only passenger comfort on this bleak windswept site. *(G.Hurst)*

Thonock Sidings Signal box is about to passed by B1 61121 (36A) on an *Up* Fish train on 18th May 1964. *(H.B.Priestley*

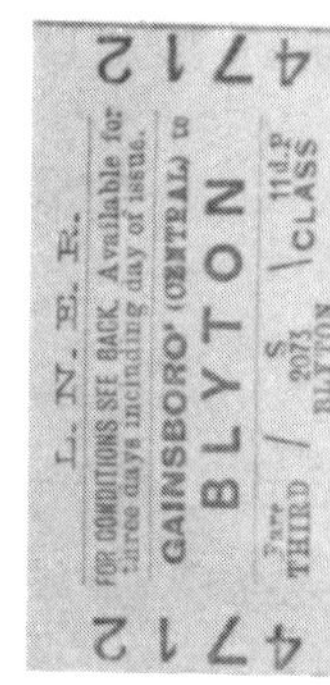

Blyton Signalbox was sited on the Gainsborough side of the station. In May 1959, the down siding has some low wagons stabled. The cattle dock and pens, and a van are located on the Corringham side of the line and approached by a road on that side. The station buildings can be seen in the background. *(D.Thompson)*

Blyton for Corringham station building as viewed from the cattle dock, on 22nd September 1956. The station had a small shelter connected to the booking office, and the platforms were split and connected by a wooden crossing. The station building built of cream brick is used as a residence. *H.B.Priestley)*

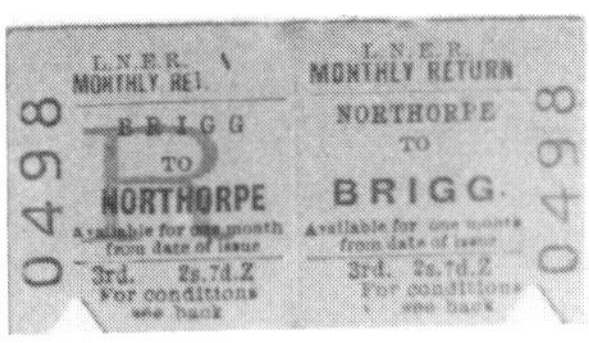

Northorpe station buildings on 30th October 1988, are partially obsured by a leylandi hedge. The gates are still in use. *(G.Hurst)*

Northorpe station on 8th August 1956, (closed the previous year) still has its enamel sign.The signal box opens and closes the gates with a wheel. Its goods depot is partially visible behind the station buildings. The signalbox survives, however the signals are colour and the line is single except for a short bidirectional loop on the Kirton Lindsey side of the station. *(H.B.Priestley)*

Kirton Lindsey Signalbox with a water barrel to collect the rainwater, is accompanied by Kirton banker N5/2 0–6–2T 69314 (36E), on 8th September 1956. It stands next to Kirton Goods depot, the last large MSL goods depot standing, along this line. *(H.B.Priestley)*

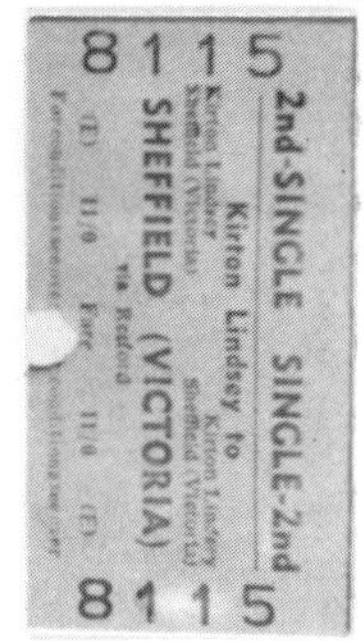

A view of Kirton Lindsey station looking east towards the tunnel at the turn of the century. The up platform has been extensively extended. Wagons full of barrels sit in the siding next to the station. Oil lamps provide the illumination for the station platform. On the down platform an awning and shelter was provided, these survived until the line was singled. (Lens of Sutton

Kirton Lindsey station on 30th October 1988, the line has been singled and the down buildings and footbridge were quickly removed. The station has been converted into a domestic residence. Half a bus shelter has been provided giving little protection against the elements. Although electric lighting is provided, in December 1988 the station had no lights. *(G.Hurst)*

Kirton Lindsey tunnel west portal was built with two brick turrets that stand high above the line. On 1st November 1988 only a single continuous welded line appears out from the centre of the tunnel. *(G.Hurst)*

Kirton Lime Works Signal Box on 1st November 1988. It is the last Great Central high level signalbox, built this way so the signalman could view the line to Scawby over the bridge parapet. Today the box controls access to the Lime Works and marks the eastern end of the single line section from Northorpe. *(G.Hurst)*

Scawby and Hibaldstow station in May 1959 looking east towards Brigg. The 'up' platform has a brick built waiting shelter. The main buildings house the booking office and small waiting shelter and the stationmasters house. The main building survives as domestic residence, but the platforms have been removed and no sign of the up platform exsists. *(D.Thompson)*

Brigg Siding signalbox was of a low level construction and controlled access to Brigg Sugar Works. Because of the seasonal nature the box was often switched out of use. Although the rail connection was severed the building survived for a couple of years before being demolished. *(M.A. King)*

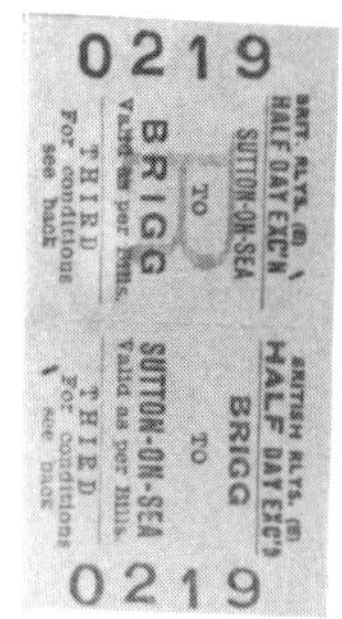

Brigg at the turn of the century looking eastwards. The station roof of the original design. On the platform stands a bookstall (a sign of an important station) and two parcel and luggage weighing machines. In the background the goods depot stands on the upside of the line. *(Heyday Publishing)*

Brigg station front was a fine two storey brick building, this view taken on 7th May 1972, shows the station entrance built with as a three arched portico. Today this building has been greatly reduced to the former stationmasters living quarters, which is now of domestic use. *(M.A. King)*

Five Mark 1's and B1 61096 (41A) stand beneath the modified station roof at Brigg, with the 12.43 pm Sheffield Victoria to Cleethorpes on 9th September 1959. *(H.B.Priestley)*

On 1st November 1988, 31203 passes through todays Brigg station, with a westward engineers empty ballast train from the Barnetby direction. *(G.Hurst)*

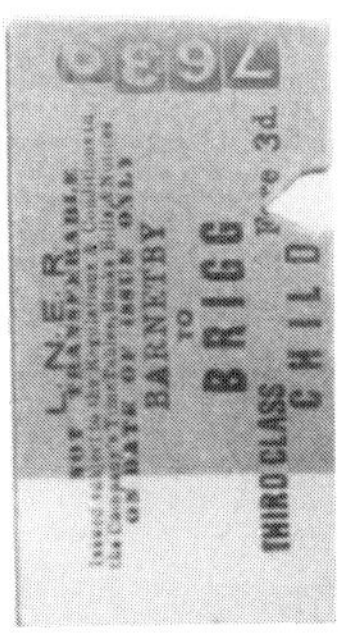

9F 92196 (40B) is about to pass Brigg Signal Box with an 'UP' fish train on 18th May 1964. *(H.B.Priestley)*

3. Clarborough Jn – Sykes Jn

The branch from Clarborough Junction to Sykes Junction was included in the original plans of the Sheffield and Lincolnshire Junction Railway, submitted in November 1844. Originally the plan was to terminate to the north of St Marks Church, Lincoln, however the Great Northern Railway (GN) had plans to build a line from London to York through Lincoln, and the line north of Lincoln also formed part of S&LJ line. Parliament suggested they only build one line.

The Sheffield and Lincolnshire Extension Act of 3rd August 1846, authorised :-

"The building of a railway from the proposed Sheffield and Lincolnshire Junction to Lincoln. It was able to raise capital of £250,000. The railway to commence in the Parish of North Leverton in the County of Nottingham and terminate in the Parish of St Marks in the City of Lincoln. Connection with Sheffield and Lincolnshire Junction. May be necessary to connect with Midland Railway Nottingham and Lincoln Railway.

The Great Northern Railways London to York Act provides between Saxilby and Lincoln identical line. Expedient for one railway. Incorporate a junction with Great Northern. Any difference to arbitration.

Level Crossings were authorised in the following Parishes -; North Leverton 1, South Leverton 3, Cottam in the Parish of South Leverton 1, Torksey 1 ,and Saxilby 2.

Power to incorporate with Sheffield and Lincolnshire Junction Railway".

The half yearly meeting held on 18th August 1847, reported that the Retford to Lincoln line was staked out.

The '*Sheffield Independent*' reported on 19th February 1848 :-"Retford to Lincoln railway, nothing to be done, but work should proceed in the spring". On 13rd April, the MSL came to an agreement with the GN over the location of the junction of the Leverton branch and the GN Gainsborough branch. Arrangements also covered the MSL use of Lincoln GN station, and the GN use of Gainsborough and Retford MSL stations. The MSL board sanctioned the agreement on 18th May 1848.

Purchase of the land from Clarborough to Hyde was deferred by the board on 3rd June, and they asked the company secretary to ascertain from Mr Fowler, whether Messrs Peto & Betts have put in a tender for the line, and at what price. The solicitors were told on 22nd June to purchase the land for the Leverton branch. Tenders had been received the same day for the branch rails and sleepers. Mr Fowler commented to the board meeting of 18th July :- "On the Leverton branch, the only major work was the bridge over the Trent at Torksey, and unless work is started on the coffer dam before the winter floods, then work will have to postponed until next spring. The GN were meanwhile pressing the MSL for early completion, and will consider any postponement as breach of the agreement".

A tender from Peto & Betts of £60,000 was received on 8th July, and at the next board meeting ten days later, it was accepted. Payments to be made monthly, half in cash and half in debentures, and to be completed by 1st April 1849.

Capt. J.L.A.Simmons RE Inspector for Railway Commissioners visited the branch on 21st December 1849 and gave his report as follows :- "The engineering works are generally light, the only one of any magnitude being a viaduct by which the railway is carried across the River Trent: this is near 400 yards in length, and consists of a number of timber–bays with wooden piles for supports, and two openings of 130 feet in each, spanned but wrought iron tubular girders resting on masonry piers. The roadway over the openings is carried upon wrought iron cross bearers rivetted to the outside principal girders, which form the parapets of the bridge, and are thus made to carry both lines of way. I tested these girders by bringing on one opening four engines in steam with their tenders (two to each line), together amounting to about 80 tons, found the deflections of each to be nearly 1¼ inch. The beams or tubes themselves appear to have sufficient strength to support the weights that may come upon them in practice, but are not built in a very accurate line, nor very regular in form, and being painted, I was not able to examine closely the rivetting; but I nevertheless remarked several rivets that appeared not to be very perfect; the deflection also is more than might, I think have been anticipated from the effect of the load, and considering the very light construction of the bridge, the superstructure over each span being only 180 tons, whereas the load that may come upon it in the event of two trains passing each other on one bay bears a large proportion to this the permanent load or weight of the bridge, and considering also the amount of deflection under the test applied, I do not feel sufficient confidence in the construction to recommend the commissioners to authorise the opening of this line for the conveyance of the public unless some means be contrived for the stiffening this bridge. In wrought iron tubular girders, great care appears to be necessary in the arrangement of the details of construction, and the departure from proportions fixed carefully by experiment should require a fresh series if experiments to arrive accurately at the knowledge of the strength of the construction, and considering the great number of parts of which a tube is formed, and the strains to which it is subject, with the probable reduction of strength, although slight, by time, I cannot do otherwise than report that, according to the knowledge as yet obtained of this subject, I do not consider that the bridge can be submitted to the continuous passage of trains for an unlimited number of times with safety, and therefore I have to report, that the opening of this railway for the conveyance of the public will, in my opinion, be attended with danger to the public using the same, by reason of the insufficiency of the works".

The Office of Commissioners of Railways informed the Manchester, Sheffield and Lincolnshire railway on 27th December that it could not open the Torksey branch for one month. Later a letter was received from Capt.J.K.L.A. Simmons, saying that he had miscalculated the weight of the engines, which should be 148 and not 80 tons. However after carefully reconsidering the calculations, he still concluded that the railway bridge at at Torksey was unsafe. He reinspected Torksey Viaduct on 11th January 1850 and tested the following :- "1st.

Three engines, the weights of which were given by the locomotive Superintendent as amounting to 108.075 tons, when fully loaded, were placed upon the south road, and produced a deflection in the south beam of .6 of an inch, and of the north beam of .36 of an inch. Three more engines, together weighing, with their loads 114.6125 tons, were then placed upon the north road, which increased the deflection of the south beam to 1.20 inch, and of the north beam to 1.32 inch. These deflections were entirely due to the temporary load. Calculations made from drawings and dimensions gave a deflection of 1.41 inch for an evenly distributed load over the whole bridge of 222.6875 tons, this being the stated weight of the six loaded engines. As I consider this experiment to be a verification of my calculations". Capt. Simmons stated that it was his opinion that the viaduct was still unable to bear the weight of two trains and refused consent to open to passengers. A letter was subsequently sent the following week informing the MSL, that no passenger services could operate for another month.

On 25th January, Mr Fowler wrote to the Commissioners saying :- "That Captain Simmons opinion was that a deflection of 1.26 inch with 222 tons is too great. He would be glad to be informed what the amount of deflection would indicate a sufficient strength. Mr Fowler pointed out that the Commissioners are now using the Girder Commissioners' Report as the new standard. But pointed out that opinion greatly differed, and that any doubts should be removed as soon as possible. He understood that Capt. Simmons used the multiple of 6 for cast iron, as the sum of the girders and roadway and the greatest passing load, and the breaking weight. He asked as this question is so wide and important that he hoped that this matter should be promptly and carefully considered and settled".

Meanwhile the Eastern Executive Committee met on 26th January, and a letter dated 18th January was read. It said that it was legal to run goods trains, but not to carry passengers. The committee ordered that goods should start immediately. Great Northern minutes reported that goods traffic commenced on 20th January, and the company was to make arrangements to commence its own goods service.

This prompted the Railway Commissioners to write to Lord Wrottesley for assistance on the meaning of certain paragraphs concerning bridges of cast iron. At the same time Mr Fowler had written to Mr Fairburn about the strength of iron bridges. Mr Fairburn replied said that he assumed Capt. Simmons had mistakes in his calculations and that his tubular bridge would need 1,300 tons equally loaded on the bridge before it would break, and that the experiments undertaken at Portsmouth did not apply to his tubular iron bridge. This letter was passed onto the Railway Commissioners who passed it onto Capt. Simmons. His opinion did not change.

Captain Laffin inspected Torksey Viaduct on 16th March and said that the condition is the same as Capt. Simmonds visit of 20th February, and after careful inspection of the structure, he concurred that the viaduct was still a danger to the public. This report was passed to Mr Fowler, by the Company secretary Mr Humphrey. Mr Fowler wrote immediately to the Railway Commisioners on 20th March :-

"I have already complained of the want of reasonable notice for which for the second time prevented my own attendance, or proper arrangements being made by the Company for the reception and convenience of the inspecting officer. The report of Captain Laffin being really but a confirmation of Captain Simmons' former reports, without any fresh reasons being assigned, and made after a cursory inspection, without trial, I shall consider as having no value beyond having ascertained, that no alteration be made".

Mr Fowler then gave his opinion on the Simmons report, and at the end of the letter wrote :-

"Under these circumstances I feel it impossible to recommend the Railway Company to make any addition whatsoever to the strength of the bridge".

Captain Simmons wrote a lengthy report on Mr Fowlers comments and near to the end said :- "I am induced to recommend that the Company be permitted to use this bridge for public traffic, provided their engineer will make such an arrangement of the platform that the ballast cannot be allowed to accumulate beyond the depth of 2 inches, upon which consideration was his calculation of the weight of the structure; and also that careful tests should be applied from time to time, with occasional inspections by an officer of this department".

The MSL board resolved on 16th April to agree with the conditions as required, and on 23rd April Capt. Laffan inspected the bridge at Torksey and reported the following :-

"I, on this occasion confined myself to ascertaining that the recommendation as to the ballast therein referred to had been complied with. I have to report that the Company's engineer has caused the longitudinal planking to be taken up and transverse planking thus comes within 3¼ inches of the top of the rails, and thus precludes the use of greater depth of ballast than a couple of inches".

Captain Simmons wrote to the Secretary of the MSL on 25th April, :-"That the Commissioners now have no objection to the line being opened for the purposes of public use".

The '*Doncaster Gazette*' said on 26th April :- " Government has at last consented to report in favour of the Leverton and Saxilby branch. Some eminent engineers have confirmed the safety of the bridge which has carried enormous test loads. Messrs Peto & Betts have been ordered to have everything in readiness for the opening in May". The Great Northern made an application to the MSL, asking whether the branch had been passed for traffic, as a delay of opening is of serious consequences to the company. Again the GN wrote to the MSL on 21st May, asking the MSL to fix a date when passenger traffic may commence over the Clarborough branch. Meanwhile the MSL wrote to the GN on 13th May saying that the agreement may be abandoned, and on 24th the company gave notice to the GN to terminate the agreement, however Lord Yarborough urged settlement of the difference.

The Great Northern took the MSL to Chancery Court in April, over its refusal to allow it to use the Clarborough branch. The MSL minutes of 20th July, recorded that the GN had won there case, and that sequestration would issue against the company, if they did not immediately admit the GN company to the use of the line, or make terms of agreement with them. They instructed Mr Fowler to arrange with Mr Alport accordingly. As a result of the courts judgement, the MSL informed the GN that it would commence operating a passenger service from

Sheffield to Lincoln GN station from 1st August, and the GN minuted on 30th July that will start there Doncaster to London service from 7th August. The GN also recorded that steps to be taken to enforce MSL compliance of the terms of the agreement between the two companies so that GN trains are not impeded.

The Board said in its half year report to shareholders that the Leverton Branch had opened to passengers on 7th August 1850. The Great Northern started services straight away with 4 trains from London to York running non stop from Saxilby to Retford.(See Table 1) Great Northern Timetable Bradshaws September 1850.

Table 1

Weekdays	A	B	B	B	B	Sunday B
	am	pm	pm	pm	pm	am
LINCOLN	918	1252	147	435	815	936
SAXILBY	934		2 2	450	829	
RETFORD	10 2	1 2	232	520	9 0	1013

Weekdays	C	D	D	D	D	D	Sunday D
	am	am	pm	pm	pm	pm	pm
RETFORD	7 9	1020	1222	232	543	923	1 2
SAXILBY	735	1048	1252	3 0			132
LINCOLN	750	11 5	110	317	616	955	2 0

A Peterborough–York, B London–York
C Doncaster–London, D York–London

The MSL local passenger services did not appear in the Bradshaw Timetables until January 1851, and are given in Table 2.

Table 2

Weekdays	am	am	pm	pm	pm	pm	pm	Sunday am	Sunday pm	Fares 1st	2nd	3rd
RETFORD	10 0	1035	2 0	232	543	647	933	10 0	725			
LEVERTON	1013			245		7 0		1013	738	*12/7*	*9/5*	*5/10¼*
COTTAM	1020			252		7 7		1020	745	*13/1*	*9/9*	*6/0½*
TORKSEY	1023			255		710		1023	748	*13/3*	*9/11*	*6/2*
LINCOLN	1040	1115	230	312	613	727	10 5	1040	8 5	*18/6*	*13/0*	*9/1½*

Weekdays	am	am	pm	pm	pm	pm	Sunday am	Sunday pm
LINCOLN	730	910	1235	140	440	936	8 0	545
TORKSEY	747		1252		457		817	6 2
COTTAM	750		1255		5 0		820	6 5
LEVERTON	757		1 2		5 7		827	612
RETFORD	810	950	115	225	520	1010	840	625

The MSL board were told on 24th February 1852, that Leverton station would cost £320 and 4 level crossing houses £150 each to construct, a decision on was postponed. It was not until 13th May that the company decided to order permanent stations for the Leverton branch.

Dwelling houses for Cottam and Leverton stations, costed in February 1852, were not approved until 3rd February 1854.

With the opening of the direct Towns line in July 1852 the Great Northern shifted its London trains over the direct route. Immediately upon opening of the Great Northern line from Doncaster to Gainsborough, on 15th July 1867, the GN diverted most of its coal trains over it and away from the Leverton branch. However the GN did not allow its powers to lapse for goods traffic until 30th June 1868, and passenger trains on 31st July.

Mr Sacré recommended to the board on 23rd November 1877, that the tender for repairs from Messrs. Handysides for £5,998 should be accepted, as it was absolutely necessary to renew the Torksey bridge due to its condition. Mr Underwood asked the board for better accommodation for Leverton on 2nd October 1883, the board decided that any expenditure should not exceed £200. Mr Underwood came back to board on 23rd January 1884 and said that £200 would not accomplish effectively, what was required, the board ordered that estimates were to be prepared for the next meeting. The board met next on 8th February, Mr Underwood estimated that additions at Leverton would cost £800, the board could not deal with this, with satisfaction at less cost. The matter was referred to Col. Hutton to report. Three weeks later, he reported that in his opinion the planned accommodation is much needed, and was prepared to recommend the same be carried out, which would involve an extra £600. Estimates were prepared for block telegraph for the Leverton branch on 16th May 1884, figures were given for Clarborough Junction to Torksey £1,478 and £1,083 for Leverton to Torksey.

On 29th March 1889 the board received an estimate for a new station, goods and cattle station at Cottam for £2,464 15s 0d, which was much required.

Part of the MSL Act of 6th July 1895 authorised :-

"A railway No.11 21.66 chains in length being a widening of the Companies Leverton Branch over River Trent, commencing in the Parish of Rampton in the County of Nottingham by a junction with that railway and terminating in the Parish of Torksey by another junction with the same railway in Parish of Torksey in the County of Lincoln".

It was not until the following year, that tenders were put out, for the erection in place, of 2 new central girders and maintenance for 6 months. The board minuted that they accepted the tender from Messrs. J. Cochrane & Sons on 15th July 1896, of £11,008 16s 0d for strengthening the Viaduct and £1,578 13s 2d for widening of viaduct. The work to be completed within 12 months.

The board were informed on 12th March 1897, that the present sand supply for locomotive purposes at Barnby Dun was nearly exhausted, however a supply had been found at Torksey. It would cost 4d per ton for next 10 years, a minimum charge of £100 per year for the siding plus a cost of £300 to join access to land.

Messrs.Moss and Sons approached the Great Central to use the sidings at Cottam for 20 wagons, for the construction of a mental asylum at Rampton. The board were told on 26th November 1909 that it would cost £879 to improve the sidings. The company to pay, but to allow a deduction of 10 percent of outlay. The existing accommodation inadequate and require improvements for goods traffic. The improvements were not provided as the board were told on 10th March 1911, that the contract for Rampton had not gone to Messrs Moss but to Messrs Pattinson & Son. They sanctioned the expenditure of £755 for an additional siding on the recommendation of the Traffic Officers, which would serve Messrs Pattinson plus Boring Syndicate for Coal who have deposited large quantities of material on the site.

The board decided that a house of two up and two down was inadequate for a married man and a family of five. Expenditure of £199 was approved by the board on 12th January 1912, to improve Leverton station.

Under LNER ownership little work appears to been undertaken. At Torksey the Works Committee on 28th June 1946 approved the simplification of the track layout, with the diamond crossing being removed and signal alterations. It was to cost £922 less £5 for annual maintenance.

With the begining of the Winter timetable in September 1956, Derby lightweights began to appear on the local passenger services, e.g. 1.14 p.m. Boston – Sheffield Victoria.

The Eastern Region announced in February 1959 its intention to close completely, the 8 miles 15 chains of direct line from Clarborough Junction to Sykes Junction. The reasons given, are that large losses were being incurred, and the extensive and costly repairs necessary on Torksey Viaduct. The previous year the line was singled and the pilotman worked between Torksey to Cottam with the down line been used to inspect Torksey Viaduct, and undertake repairs. Electric token was brought into use and a facing crossover installed at Torksey. A 25 mph speed restriction applied over this stretch of track.

The 2nd November 1959 saw the last day of passenger operation between Retford and Lincoln Central over the direct Leverton line.

Table 3

Weekdays	am	am	am	pm	pm	pm	pm	pm	pm	Sunday am	Sunday am	Sunday pm	Sunday pm	Sunday pm
LINCOLN C.	735	920	1140	12 0	238	4 0	538	643	918	944	1054	2 0	518	720
SAXILBY	744	929		12 9	247	4 9	547	653	927		11 3	2 9	527	729
TORKSEY	751	936		1216	254	416		7 0	934		1110	216	534	736
COTTAM	756	941		1221	259	421		7 5	939		1115		539	741
LEVERTON	8 1	946		1226	3 4	426		714			1120		544	746
RETFORD	813	950	NCC	1238	316	443	618	726	954	1020	1132	238	556	758

Weekdays	am	am	am	am	am	pm	pm	pm	pm	pm	pm	Sunday am	Sunday pm	Sunday pm	Sunday pm
RETFORD	430	711	815	9 5	11 0	2 5	NCC	429	6 0	718	9 2	1055	1252	622	855
LEVERTON		722		916	1111	216		440	611	729	913	11 6			9 7
COTTAM		726		920	1115	220		444	615	733	917	1110			911
TORKSEY		732		926	1121	226		450	621	739	923	1117	112	642	917
SAXILBY		738	845	932	1127	232		458	627	745	929	1122	119		924
LINCOLN C	5 8	748	854	942	1137	242	445	5 8	637	755	939	1132	129	659	934

NCC The North Country Continental

The Gainsborough Model Railway Society requested that the last regular trains should be steam hauled instead of the normal dmus. The 7.20 p.m. from Lincoln and 8.55 p.m. return from Retford were hauled by J11/3 64318 (40A) and six coaches. The '*Railway Observer*' reported :- "Engine was spotless, with cleaned brass beading on the splashers, the smokebox and front end had been painted black, and the buffer beam and vacuum pipe red". The end was seen off by large numbers of passengers, however the last train over the line was a return excursion from Manchester to Boston hauled by Lincoln B1 61248 'GEOFFREY GIBBS', which should have departed from Retford at 9.53p.m., but due to late running departed from Retford at 10.28p.m. without any crowds.

From Monday 3rd November all passenger and goods trains were diverted via Gainsborough Lea Road adding 25 minutes to the average journey. Although the line totally closed, the track was left insitu, with a planned coal fired power station at Cottam. By June 1962 the access at Sykes Junction had been reduced to single facing points. A short spur leaving the line at milepost 71½, was constructed to the power station, with a loop for the new generation of freight wagon for the revolutionary Merry Go Round (MGR) system.A fixed formation of wagons straight from the colliery to the power station. Cottam power station started generating in June 1968. Signalling of the line, now being controlled from Thrumpton (Retford) Signal Box. Coal is supplied from the North Derbyshire, Nottinghamshire and South Yorkshire collieries. Fuel oil trains were booked twice a week, on Tuesday and Thursday from Lindsey to Cottam Power Station. These trains have to travel to Worksop Yard, in order to run round. However for the last two years these oil trains usually are condensed into 5 trains (one per day for a week) every 3 or 4 months.

The section from Sykes Junction to Torksey lay disused until 31st January 1966, when Shell opened an oil terminal at Torksey. The oil trains ran over the now single track line from Torksey to Stanlow Oil Refinery thrice a week on Tuesday, Wednesday and Friday. Traffic ceased by Summer 1988 and Sykes Junction Signal Box was taken out of use on 18th September 1988,and the signal box was soon demolished. The pointwork at Sykes Junction was removed on Sunday 23rd March 1989 and replaced by plain track.

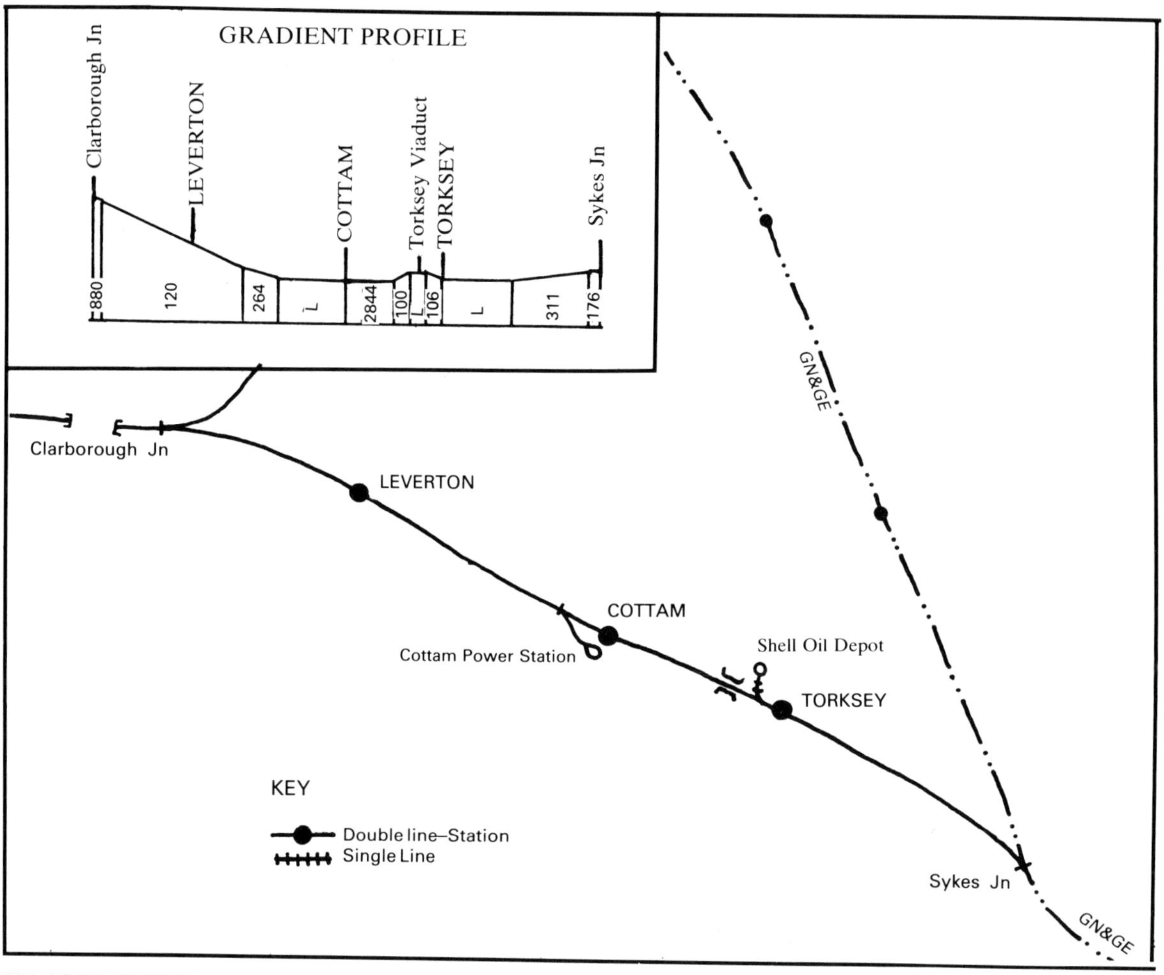

Clarborough Junction Signal Box on 7th August 1964, looking east showing junction to Leverton. The signal box was the only brick built one alone this line, of unknown origin. *(H.B. Priestley)*

A large crowd stands on Leverton station at the turn of the century, awaiting the arrival of a Retford bound train. The train is approaching in the distance and the signalman has closed the gates and signalled the train to enter. Today the site has been flattened and nothing remains. *(Heyday Publishing)*

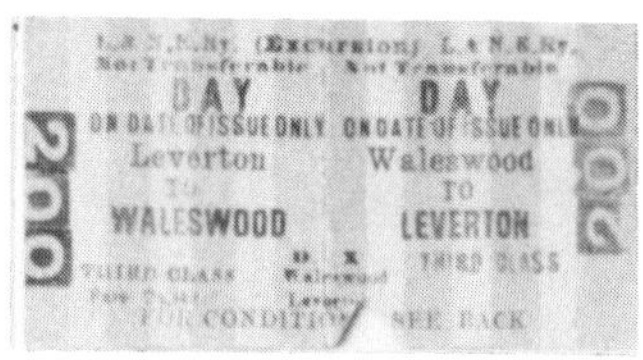

(above) The up platform of Leverton station in May 1953. The lighting is provided by oil lamps. A brick waiting shelter was added to improve passenger accomodation. *(D. Thompson)*

(left) Cottam Power station cooling towers loom large over 56020 on 31st March 1989, as it haules its train of empty MGR wagons back to the colliery. *(Geoff Hurst)*

(above) Cottam station circa 1900, looking east towards Torksey. The station has two waiting shelters, one of wood on the 'up' and brick on the'down' platform. In the background wagons fill the goods yard and indicates a heathly freight traffic. *(G.Hurst Collection)*

The fireman of 02/3 63961 (36E) on a down goods train, comes to a slow to accept the token from Cottam signalman, and cross over onto the 'up' and run wrong line to Torksey, on the temporarly singled line over Torksey Viaduct on 30th October 1959. The last week of operation. *(H.B.Priestley)*

A view taken from the trackbed of Cottam station on 31st March 1989, the stationmasters house has survived as a domestic residence. The extension crosses onto the old trackbed. Surprisingly the wooden waiting shelters remain (hidden in the scrub) and out of sight, the brick waiting shelter stands after nearly thirty years of disuse. *(G.Hurst)*

Torksey Viaduct nearly 30 years after closure. The later addition of the strengthening beam rises over the original tubular iron structure. The line on the Torksey side of the bridge is carried on an iron tressle. *(Geoff Hurst)*

The Shell Oil terminal was the only source of traffic on the Torksey branch. It was sited next to the River Trent. The depot acted as the loading point some of oil field of Lincolnshire and Nottinghamshire. On 9th February 1988, 47197 backs part of train into the second loading bay. When loaded the train will travel to Stanlow (via Doncaster, Healey Mills and Manchester) for refining. *(A.Morgan)*

Torksey signal box 'up' home signals the approach of a train and two gentlemen walk along the platform in July 1953. The booking office is the single storey building next to the passenger crossing. (D.Thompson)

(middle) A view of four members of staff at Torksey station, pose for this turn of the century view. *(Lens of Sutton)*

(below) 47278 on 11th August 1987 heads a Torksey – Stanlow oil tank train. The signalman having just received the token for the Torksey branch. *(A.Morgan)*

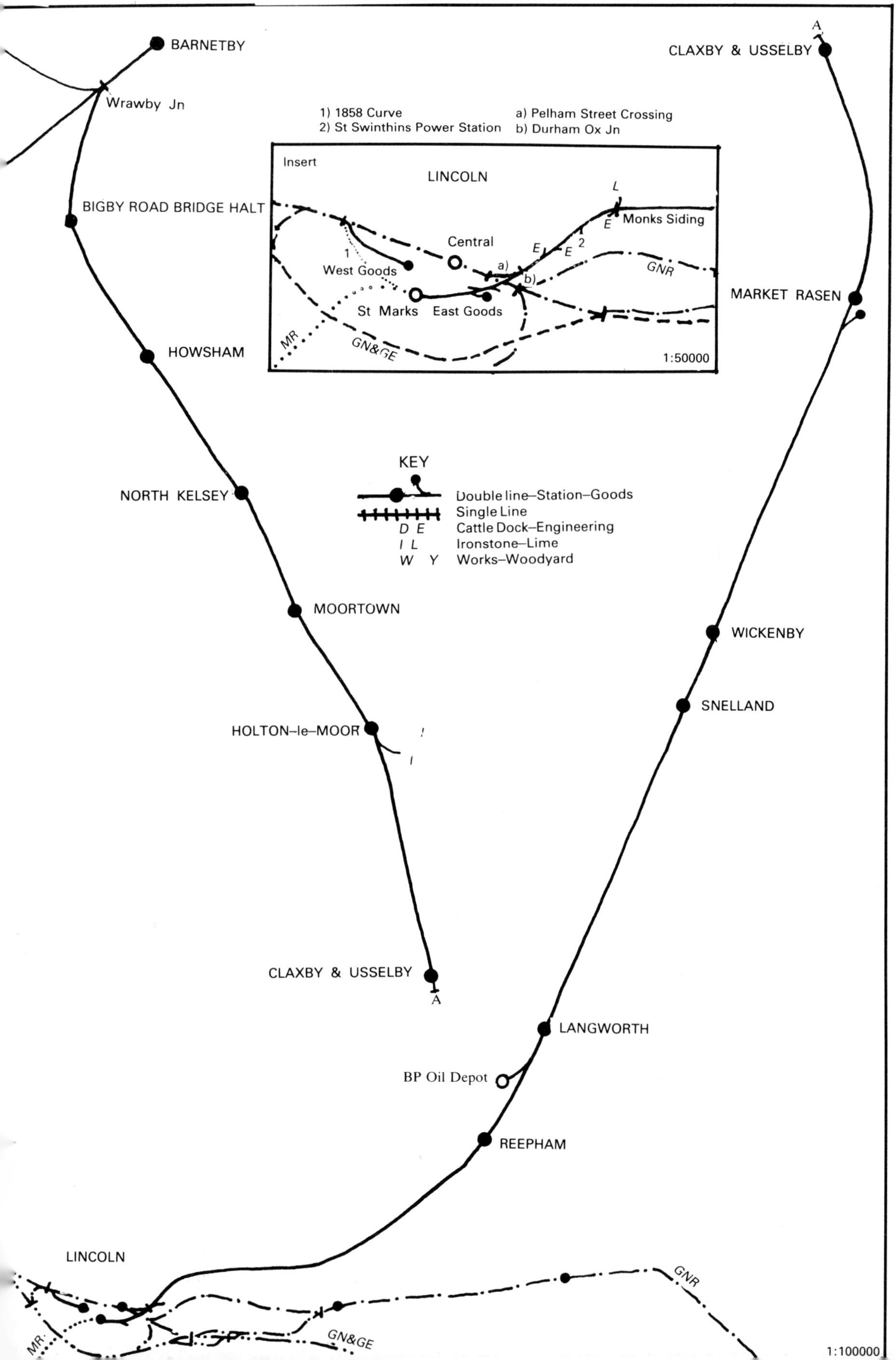

BARNETBY
Wrawby Jn
BIGBY ROAD BRIDGE HALT
HOWSHAM
NORTH KELSEY
MOORTOWN
HOLTON–le–MOOR
CLAXBY & USSELBY
A
CLAXBY & USSELBY
MARKET RASEN
WICKENBY
SNELLAND
LANGWORTH
BP Oil Depot
REEPHAM
LINCOLN
GNR
MR
GN&GE
1:100000
1) 1858 Curve
2) St Swinthins Power Station
a) Pelham Street Crossing
b) Durham Ox Jn
Insert
LINCOLN
L
E
Monks Siding
Central
West Goods
St Marks
East Goods
1:50000
KEY
Double line–Station–Goods
Single Line
D E Cattle Dock–Engineering
I L Ironstone–Lime
W Y Works–Woodyard

4. Lincoln – Wrawby Jn

The newly formed Great Grimsby and Sheffield Junction railway submitted plans in November 1844, for a mainline from Grimsby to Bole and included a branch line of 14 miles from Bigby to Market Rasen. The junction for the branch faced west towards Brigg.

The Market Rasen branch was authorised as part of the Great Grimsby and Sheffield Junction Act of 30th June 1845. This act sanctioned the building of railway, commencing in the Parish of Bigby and terminating in the Parish of Market Rasen. The Act permitted the crossing of the turnpikes, on the level in the following parishes :- Bigby 2, Cadney cum Howsham 1, North Kelsey 3, South Kelsey 1, Holton le Moor 2, Claxby 1, Usselby 1, Walesby 2 and Middle Rasen 2.

Later that year, the Great Grimsby and Sheffield Junction Railway submitted two bills. The first extending the Market Rasen branch to an end on connection with the Midland at Lincoln, and the Rasen branch was amended to include a Bigby East Curve of 1 mile 36 chains. Leaving the mainline 22 miles 10 chains from its start at Bole, and terminating 60 chains along the Bigby West Curve, which now left the mainline 20 miles 50 chains from Bole. The second proposed a branch to Caistor, which left the Market Rasen line 5 miles 50 chains from its junction and climbed for 2 miles 47 chains to terminate on the western edge of the town. The first 64 chains were to be laid at 1:160 and the rest of the line at 1:60.

The first Great Grimsby and Sheffield Junction Act of 26th June 1846 authorised the following works :- Extension of the railway from Market Rasen to Lincoln by a junction with the Midland Railway Branch. To terminate in the Parish of St Marks by a junction with the Nottingham, Newark and Lincoln Railway or by a distinct terminus. It allowed the following roads to be crossed on the level in the Parishes of Market Rasen 1; Buslingthorpe 1; Wickenby 1; Snelland 1; Stainton by Langworth 1; Scothern 1; Barlings 1; Reepham 1; Cherry Willingham 3 and St Marks 2. To erect stations or lodges at level crossings for the convenience and security of the public. All works to be completed within seven years. Navigation rights of the River Witham were protected with a clause defining the width and height of the railway bridge over that river. The company was empowered to raise upto £265,000.

The second Great Grimsby and Sheffield Junction Railway Act of 26th June 1846, enabled the company to construct a branch to Caistor and abandon and deviate from the line authorised in the 1845 Act. Increased borrowing powers to £96,000.

The GG&SJ board in November 1846 awarded the contract to construct the extension to Lincoln to Messrs Waring for £73,000.

In 1847 the Manchester, Sheffield and Lincolnshire Railway (MSL) and East Lincolnshire Railway jointly submitted plans for a Lincoln, Louth and Horncastle Railway.It left the Lincoln to Market Rasen line, where the main road crossed over the line. The main line to Louth passed through the parishes of Goltho; Wragby; Market Stainton; Stenigot; Raithby cum Maltby to join the East Lincolnshire railway just south of Louth. At Withcall, a tunnel of 1,341 yards was planned to pass through the Wolds. The Horncastle branch, left the Lincoln to Louth line 7 miles 33 chains from its western junction. The branch to be 8 miles 47 chains long.

These plans were greatly reduced, and the MSL Act of 9th July 1847, authorised the construction of a branch from the Market Rasen to Lincoln railway to the town of Wrawby. Commencing in a field in the Parish of Stainton cum Langworth with Rearby and Newball in the County of Lincoln, passing through the parishes of Rusby, Reresby, Stainton Wood, Bullington, Goltho cum Bullington, Bullington to terminate in the Parish of Wragby. To erect and maintain stations or lodges, where railway crosses roads on the level. For each offence the company will pay a penalty of £20. To be constructed within seven years of passing of this act. This railway was never built.

The Manchester, Sheffield and Lincolnshire Board awarded the Market Rasen contract on 6th January 1847 to Messrs Waring for £44,986 11s 3d and was to be completed by January 1848, with a penalty of £1,000 per week. The board meeting held on 18th August, minuted that contracts for stations on the line had been let to Mr Kirk (part of the Lincolnshire lines) and work on the line was delayed by non possession of the land.

The two local papers reported construction work on the branch. On 3rd October 1847 the '*Stamford Mercury*' reported that :- "The cutting through Greetwell Hill is nearly completed. At the furthest extremity abridge has been completed carrying the Fiskerton Road over the line and beyond. The works are in a state of great forwardness. The embankment through the valley through Greetwell Hill and the Witham is also nearly completed. Preparations have been made to carry ballast over the Witham to finish the embankment where the line crosses the GN. The permanent bridge over the Witham has been commenced and also the bridge over Sincil Dyke"

The '*Stamford Mercury*' of 14th January reported :-"Preparations have been made to carrying the GG&SJ railway over the High Street into the Midland station. Houses to the east of the road have been pulled down". On 11th February the Lincolnshire Chronicle reported "Coffer dams of the bridge just below the locks at Lincoln were again flooded". The 31st March edition of the Stamford Mercury gave the following news :- "The earthworks between Lincoln and Market Rasen have been suspended in consequence of an injunction to restrain the MSL from tipping the bank on GN property". On the 24th, the MSL had resumed tipping where they had been stopped by the injunction. Tipping again stopped. The Stamford Mercury of 11th August 1848 said that :- "Market Rasen station is progressing towards completion. All the brick work in the building is finished, and the roof is on. The level crossings on the Rasen branch are in a forward state. The stations on the Rasen to Lincoln line are likewise in a forward state".

Mr Cubitt was asked by the GN board on 31st August to arrange with Mr Fowler, about the forming a junction with the MSL east of Lincoln station.

The Eastern Executive Committee visited the works on 9th, 10th and 11th November 1848 and minuted :- "Market Rasen to Lincoln, works in very satisfactory state, only impediment ballasting 2 miles south of Rasen and Witham Bridge". The *'Stamford Mercury'* of 24th November reported "The first girder of the bridge of the Grimsby and Sheffield Junction railway over River Witham was put in on Wednesday last (21st), the girders are made of wrought sheet boiler iron, and amazing strength required appears to be gained by a peculiar way in whichthey are made the deflection of such girders as these are slight indeed". While the *'York Courant'* described the Witham Bridge :- "The bridge near Lincoln is almost complete, the sides of the bridge are composed of hollow tubes of sheet iron about 6 foot high and 18 inches wide to a lower part these tubes cross girders and sleepers are attached by copper bolts, upon these sleepers two immense girders extend the entire length of the bridge from pier to pier, and on them the rails are placed so that the whole weight of sleepers, rails and traffic will be borne by the iron tubes". On the same day the Lincolnshire Chronicle stated that the wall on the west side of High Street, was already down, and the rails will soon be laid across the street. While on 24th November it said that :- "The Midland station at Lincoln is being prepared for the use of MSL trains".

The line from Barnetby to Market Rasen opened on 2nd December 1848, without any ceremony. The line to Lincoln being incomplete, an omnibus ran the final section. The first timetable from New Holland to Market Rasen plus local fares are shown in Table 1.

Table 1

Weekdays					Sunday		Fares		
	am	pm	pm	pm	am	pm			
HULL	*815*	*1215*	*315*	*515*	*815*	*515*			
NEW HOLLAND	9 0	1245	345	545	845	545			
BARNETBY	935	120	420	620	920	620	*2/6*	*1/10*	*1/4*
HOWSHAM	945	130	430	630	930	630	*3/3*	*2/5*	*1/8½*
NORTH KELSEY	950	135	435	635	935	635	*3/6*	*2/7*	*1/10*
MOORTOWN	955	140	440	640	940	640	*3/9*	*2/9½*	*1/11½*
HOLTON	10 0	145	445	645	945	645	*4/0*	*2/11½*	*2/1*
USSELBY	10 8	153	453	653	953	653	*4/6*	*3/4*	*2/4*
MARKET RASEN	1015	2 0	5 0	7 0	10 0	7 0	*4/11*	*3/8*	*2/6½*

Weekdays					Sunday		Fares		
	am	pm	pm	pm	am	pm	1st	2nd	3rd
MARKET RASEN	743	1243	343	634	1058	643			
USSELBY	750	1250	350	650	11 5	650	*0/5*	*0/4*	*0/2½*
HOLTON	758	1258	358	658	1113	658	*0/11*	*0/8½*	*0/5½*
MOORTOWN	8 3	1 3	4 3	7 3	1118	7 3	*1/2*	*0/10½*	*0/7*
NORTH KELSEY	8 8	1 8	4 8	7 8	1123	7 8	*1/5*	*1/1*	*0/8½*
HOWSHAM	813	113	413	713	1128	713	*1/8*	*1/3*	*0/10*
BARNETBY	823	123	423	723	1138	723	*2/5*	*1/10*	*1/2½*
NEW HOLLAND	856	156	456	756	1211	756	*4/5*	*3/4*	*2/2½*
HULL	*926*	*226*	*526*	*826*	*1241*	*826*	*4/11*	*3/8*	*2/6½*

The following week after the opening of the Market Rasen branch the *'Stamford Mercury'* announced that :- "The Great Northern, after delay of some weeks have recommenced works on a curve, which is to join the line with the Grimsby and Sheffield in Humble Carr and they have nearly affected a junction". MSL minutes of 14th December said that plans of the junction with GN east of the Lincoln station were referred to the Midland Railway. The Friday 18th December edition, reported :- "The bridge of Sheffield and Grimsby Railway over Witham was subjected to test of having a weight of 150 tons drawn over it, the deflection of the girders was very trivial".

The *'Stamford Mercury'* reported that in Market Rasen :-"Great preparations are being made for celebration of opening of the railroad between this place and Lincoln on Monday 18th. A grand dinner and ball are advertised to take place in the Gordon Arms Inn, and will be very distinguished patronage. There will also be a dancing at White Hart Inn; for a million. The ball at the Gordon Arms Inn advertised "1 pint of wine and dessert for 5 shillings!"

The GG&SJ Lincoln extension railway was inspected on 14th December by Lt. General Paisley and authorised its use. The line opened on 18th December 1848 without any ceremony. The *'Stamford Mercury'* described the Barnetby line :- "The first train left Lincoln for Hull at 7.15a.m. and reached New Holland Ferry at 9.45a.m. and Hull Pier at 10.15a.m. The distance from Lincoln is 45½ miles. The line starts from the Midland Station at Lincoln, crosses the river Witham (over which a handsome bridge, on the tubular principle, has been erected), and runs close past the ruins of the Monks' Abbey; it then proceeds through Willingham, Reepham, Langworth and Wickenby, to Rasen; from Wickenby for 15 miles north, the land on either side the line is poor, and the prospect uninviting, being only relieved now and then by views of the Wolds on the east. The event was celebrated at Lincoln by a dinner at the Durham Ox Inn, at which the Mayor presided. The Bluecoat boys were treated with a ride to Hull". Bradshaws Timetables printed a timetable for the Lincoln to Hull service one month in advance. Table 2 gives the timetable and fares to and from Lincoln for December 1848.

Table 2

Weekdays					Sunday		Fares		
	am	am	pm	pm	am	pm	1st	2nd	3rd
HULL	*815*	*1130*	*230*	*515*	*815*	*515*			
NEW HOLLAND	845	1200	3 0	545	845	545			
GREAT GRIMSBY *	*749*			*532*					
BARNETBY	930	1230	330	615	930	615	*2s6d*	*1/0*	*0/11½*
HOWSHAM	940	1240	340	625	940	625	*3/3*	*2/5*	*1/8½*
NORTH KELSEY	946	1246	346	631	946	631	*3/6*	*2/7*	*1/10*
MOORTOWN	952	1252	352	637	952	637	*3/9*	*2/9½*	*1/11½*
HOLTON	956	1256	356	641	956	641	*4/0*	*2/11½*	*2/1*
USSELBY	10 4	1 4	4 4	647	10 4	647	*4/6*	*3/4*	*2/4*
MARKET RASEN	1012	112	412	657	1012	657	*4/11*	*3/8*	*2/6½*
WICKENBY	1022	122	422	7 7	1022	7 7	*5/7*	*4/2*	*2/11½*
SNELLAND	1027	127	427	712	1027	712	*5/10*	*4/4*	*3/0*
LANGWORTH	1035	135	435	720	1035	720	*6/0*	*4/9*	*3/3½*
REEPHAM	1040	140	440	725	1040	725	*6/7*	*4/11*	*3/4½*
LINCOLN	1052	152	452	737	1052	737	*7/0*	*5/6*	*3/9½*

Weekdays					Sunday		Fares		
	am	pm	pm	pm	am	pm	1st	2nd	3rd
LINCOLN	715	12 0	230	6 0	1015	6 0			
REEPHAM	725	1210	240	610	1025	610	*0/10*	*0/8*	*0/5*
LANGWORTH	730	1215	245	615	1030	615	*1/1*	*0/10*	*0/6*
SNELLAND	738	1223	253	623	1038	623	*1/7*	*1/3*	*0/9½*
WICKENBY	743	1228	258	628	1043	628	*1/10*	*1/4*	*0/11*
MARKET RASEN	753	1238	3 8	638	1053	638	*2/6*	*1/10*	*1/3*
USSELBY	8 1	1246	316	646	11 1	646	*2/11*	*2/3*	*1/5½*
HOLTON	8 8	1253	323	653	11 8	653	*3/5*	*2/7*	*1/8½*
MOORTOWN	813	1258	328	658	1113	658	*3/8*	*2/9*	*1/10*
NORTH KELSEY	818	1 3	333	7 3	1118	7 3	*3/11*	*2/11½*	*1/11½*
HOWSHAM	824	1 9	339	7 9	1124	7 9	*4/2*	*3/2*	*2/1*
BARNETBY	834	119	349	719	1134	719	*4/11*	*3/9*	*2/5½*
GREAT GRIMSBY *	929	210		812		812	*7/2*	*5/4*	*3/7*
NEW HOLLAND	9 7	152	432	752	12 7	752	*7/0*	*5/2*	*3/5½*
HULL	930	222	5 0	856	1240	835	*7/6*	*5/6*	*3/9½*

* Change Ulceby

On 31st December, Mr Cubitt of the GN, was told to arrange a meeting with Mr Fowler of the MSL. About forming a junction with MSL east of the GN Lincoln station. It was not until three months later on 12th March 1849, that Mr Cubitt suggested to the MSL to form a junction by means of a third line across Pelham Street. The Great Northern said it had the right to do so, using powers in the GN – MSL agreements. The MSL board told Mr Fowler on 4th April, that the junction be laid, according to the plans. A single line was subsequently built, but no parliamentary act sanctioned its construction.

The Executive Committee were informed on 19th July 1849, that the approach road to North Kelsey had been damaged by contractors, and they ordered the repair work to be undertaken and charged to the contractors. The minutes of 20th February 1850 gave the final costs as follows :-Market Rasen station, platform and ironwork £5,886 0s 8d; Goods Station £3,020 2s 7d. The rest of the stations were listed as 3rd class and cost £420 each with £280 18s 8d for the platforms and fences. Level crossings and houses cost £249 1s 4d each.

Bigby Road Bridge station first appeared as a footnote in the Bradshaws Timetable for March 1852. It was served by a single train the 1.55 p.m. New Holland – Lincoln train on Thursday only (Barnetby Market Day). Although in subsequent editions the 10.5 a.m. from Lincoln stopped. It disappeared from the Bradshaws Timetables in August 1882 and was served by a New Holland to Lincoln train until closure.

The MSL Act of 26th June 1855, part of which authorised a branch railway in Lincoln. A short connecting loop, between Great Northern line at Milepost 139 in the Parish of Boultham and Midland Railway near station of same railway at Lincoln.

The board heard on 13th January 1858, that the Great Northern would prefer that the curve should not be put in. One week later the MSL decided that the curve at Lincoln should be completed.

Capt. R.M.Laffan a Board of Trade inspector visited the Lincoln Curve on 13th October 1858 gave the following report :-

"In order that Manchester, Sheffield and Lincolnshire trains from Retford and the west should connect with the Rasen branch the company constructed a branch from the Great Northern west of Lincoln onto the Midland west of its station. The Manchester, Sheffield and Lincolnshire Railway Company have purchased and enclosed the land for the Lincoln branch for a double line, but they have only laid it as a single line doubled at the junctions. It is about 57 chains in length, a great part of it level, and with no work whatever of any importance upon it. It is therefore, in my opinion, much to be regretted that the branch, situated, as it is, should not have been laid as a double line – a security in itself for safe working, which can never be given, under the best regulations, by a single line. In this case the risks are multiplied, for the branch belongs to one company, and the junctions at either end, with the pointsmen to be employed, to a second and a third company. The permanent way of the branch is in good order, and the junction and distant signals at either extremity are of a satisfactory description; but no means of communication have been provided between the pointsmen at each end of the branch. It is essential that the system under which the branch is to be worked, and the instructions which are to be given to the pointsmen, should be fully considered by the three companies concerned; and I have requested that the details, when agreed upon, may be submitted for their Lordships' consideration. The most satisfactory conclusion would doubtless to be double the line of rails; and I beg strongly to recommend to the Directors to adopt that course.

As I do not consider that the safe working of this branch is sufficiently provided for by the ordinary junction and distant signals, it is my duty to report that the opening of the Lincoln branch cannot, in my opinion, be sanctioned without danger to the public using the same."

Capt. G. Ross reinspected the branch on 30th January 1859 and found the following :- Telegraphic communication has been established between the pointsmen's boxes at the extremities of the branch, and magnetic telegraphic instruments placed in them; and the following regulations have been drawn up by the superintendent.

Rules for working the Lincoln new Junction Branch Telegraph.

1st. When a train or engine is ready to be passed over the new branch line, the signal–man at the entering junction must call the opposite junction signal–man's attention by ringing the telegraph instrument bell "*once*", which he may repeat slowly till it is acknowledged by the instrument bell being once rung in return by the receiving signal–man.

2nd. The one ring acknowledgement denotes "*wait*".

3rd. When the receipt of a signal has been acknowledged by giving "*wait*", if the main line and branch are all clear, the receiving signal–man must "*stop*" the main line, then lower the branch signals to "*caution*", and afterwards ring the telegraph bell "*three times*" slowly, as notice of permission for the waiting train or engine to be allowed to enter upon the branch.

4th. No engine must on any account be allowed to enter upon the branch till the signal–man has received the three ring signal permission at the opposite end of the branch.

Capt. G.Dalton wrote to the company secretary stating that they have received Capt. Ross (RE) report on the Lincoln branch, and as long as the company used the regulations for working the line they would have no objections to opening the line to passenger traffic.

The Great Northern strongly objected to the MSL Lincoln Curve, it appears that its function for diverting MSL trains into the Midland Station and on to the Market Rasen line never really happened. The line was hardly used and plans for 1877, show the line as lifted and only the fencing marking its line.

The MSL over the next 30 years undertook improvements to stations and provided sidings as required. Outside Lincoln the major traffic consisted of agricultural produce and iron ore out and coal and industrial goods in. Langworth Station required a gate that lead onto Regent Road, and this was approved on 9th November 1860, for an estimated cost of £1 10s. It was decided on 19th June 1863 to partition the booking office at Moortown, to provide a waiting room costing £6. The company in September 1865, asked Mr Small to make a report on the Caistor branch, and which was authorised in 1845. It appears from minutes that no report is recorded as presented.

In 1867 Ruston and Proctor asking for permission to build a siding at Lincoln, the board were informed that at present, the traffic is handled by the Great Northern and Midland railways, and since 1859 revenue from Ruston and Proctor had amounted to £2992 12s 5d. It was estimated it would cost £392, and the Board approved its construction on 7th June, to use similar agreement to one used for Clayton and Shuttleworth. Later that year the West Yorkshire Iron and Coal Company approached the MSL on 8th November to construct a siding at Holton le Moor.They said it wished to convey 20 wagons of ironstone from the Mid Lincolnshire mines to its ironworks at Ardsley. The MSL estimated that the siding would cost about £300 and that the whole cost should be borne by the company. Production started at the Acre House Mine just to the south of Holton le Moor in 1868 and was connected by tramway, to the MSL. This mine reached a maximum output of 70,000 tons in 1873, but by 1881 it had reached negligible amounts and the mine closed in 1885.

On 7th January 1876, the company decided to construct a small waiting room at Langworth for an estimated cost of £100. Ironstone mines opened in 1873 at Greetwell on the northside of the line, and by 1877 the ironstone was carted to a siding at Monks Abbey. Production continued until the 1930's when the mines were subject to flooding and in 1939 output ceased. The quarries were then sold to a limestone company who stopped any output going by rail by 1944. In 1877 extensive work was undertaken involving the bridging of Sincil Dyke to provide extra sidings, it was estimated that it would cost £15,745. Mr Swan who owned some limekilns about 2 miles from Monks Abbey approached the MSL to build a branch with 3 level crossings. It was reported on 25th October 1878 that no more could be done until Mr Swan received a reply from the ecclesiastical authorities who owned the land. The scheme did not materialise.

A project to build a tramway from Lincoln to Brigg was announced in 1880 and took everyone by surprise. The board discussed the possible working of the line and reports were asked to be made. At a meeting on 26th January 1881, director Col.Hutton commented that he does not think it will proceed. Three months later, more details had appeared, the tramway was be 20 miles 40 chains long and it was estimated to cost £199,500. The matter was deferred by the board.

Even though Pelham Street level crossing was used by the MSL and GN, it was the MSL who decided on 8th June 1888, to obtain tenders to build a pedestrian footbridge. The provision of Horse and Carriage loading accommodation for Wickenby was approved by the company on 9th November 1888, at an estimated cost of £135. A board meeting of 27th July 1889, were informed that it was estimated to cost £6,853 to install block telegraph onto the Market Rasen branch.

The board meeting of 9th October 1891, approved three projects improving staff, passenger and goods customers. They were a staffhouse at Market Rasen, a waiting room, and improvements to the Stationmasters house as well as a waiting shed at a cost of £245 at Langworth. One month later a cattle loading wharf for Usselby, was approved for an estimated £150. The company authorised the building of a grain loading wharf at North Kelsey, for an estimated £270, on 7th September 1894. This saved carters having to travel to Lincoln to unload there grain. The company undertook no more alterations to rural stations or to improve agricultural transport facilities. Usselby was renamed to Claxby and Usselby on on 1st July 1897.

The Lancashire, Derbyshire and East Coast Railway (LD&EC) opened there line to Pyewipe Junction on 16th November 1896, and used there running powers into Lincoln GN station. Later the LD&EC obtained powers to operate over the MSL line from Durham Ox Junction to Grimsby, but they did not use them until 29th July 1901.

Plans were submitted in November 1904 for the widening of the connection into the Great Northern station in Lincoln. The plans included two short branches to a proposed goods depot to the west of Lincoln.

The Great Central Railway Act of 4th August 1905 authorised :-

"Railway No.1 41.7 chains in the County Borough of Lincoln on the West Bank of Witham at a point 4½ chains from the middle of the bridge carrying Great Northern Railway over said river, along the west bank there of in a southerly direction and terminate by junction with Great Northern at a point 9 chains from middle of Holmes West Junction Signal Box measured in an easterly direction".

"Railway No.2 6.5 chains By junction with Midland Railway Goods Yard sidings to a point on northern most siding 1.75 chains from bridge over River Witham measured in westerly direction terminating by junction in field No.108".

Dock on River Witham 3.5 chains south of bridge carrying Great Northern Over River Witham.

"A widening or doubling. 4.7 chains in length on the south side of railway connecting the Great Central with Great Northern at Durham Ox Level Crossing wholly within City and County of Lincoln, commencing at a point 3 chains or thereabout eastwards from centre of road one Durham Ox Level Crossing and terminate at a point on Great Northern distant 2 chains or thereabouts westward".

After 31st March 1906, the Great Central diverted all passenger trains into Lincoln GN station, using the newly doubled curve from Durham Ox. The GC continued to serve its East goods depot, and the Midland Railway used its powers to operate into the GN Holmes Goods. Otherwise the connection with the Midland was only used by the GC to transfer goods and special passenger trains.

A 65 feet turntable for the new West Goods Yard was approved by the board on 11th January 1907, with the contract going to Ransome Roper for £460. It was noted on 18th October, that Messrs Lovatts contract for the dock and yard was complete and that the Hughes & Stirling contract for the warehouse building was erected and the roof erected. The lifting appliances and cranes are now under construction.

A 10 mega watt power station was built next to the River Witham, east of the line to Market Rasen. St Swithins Powers Station commenced generating in 1921-22. It was rail connected with coal supplies coming from the Nottinghamshire coalfield over the former LD&EC line.

The three level crossings in central Lincoln always caused operating problems. A maximum of 49 wagons was allowed on trains passing through Lincoln High Street, any more would foul the level crossings at each end. This disrupted traffic and increased operating costs. The traffic department wanted to increase loads upto 80 wagons for Immingham export coal trains.

The LNER proposed to construct a new railway of 62 chains connecting the avoiding line and crossing the Loop line. The new line would cross the Lincolnshire Loop line by a 15 feet high bridge with a span of 58 feet. The River Witham was spanned by a bridge of 100 feet long and 16 feet 6 inches high. This proposed line was the only one out of 3 projects to meet the LNER Board requirements on the level of Government subsidy.

This line was authorised as part of the LNER Act of 1st August 1924 :-

"Railway No.4. 62 chains. Parish of Canwell in the City of Lincoln to Parish of Greetwell. Commencing by junction with Lincoln and Sleaford railway . Terminate by junction with companies Market Rasen Branch."

The LNER Traffic Committee were told on 28th July 1927, that under the Act of 1924, the Greetwell Jn to Monks Abbey line powers would run out on 1st October. They decided due to serious fall in coal shipments, it was rendered unnecessary, but when it revives new facilities would be required.

Ironstone production restarted at Holton le Moor in 1934, when the Mid Lincs Iron Company commenced mining at Nettleton.The mine was connected by aerial ropeway to exchange sidings at the base of the escarpment.

The LNER and LMS came to an agreement during late 1936, that the LMS would concentrate its freight traffic in Lincoln at the LNER Goods Depot. The LNER would have to provide 6 additional dray berths to cope with the extra 237 wagons per day. It would cost £280, but the company would save £46 in overtime.

The railways were nationalised in 1948 and the LNER lines in this area passed to Eastern Region control. A through passenger train was commenced at the beginning of the winter timetable from Cleethorpes to Birmingham via Leicester. Problems with Arley Tunnel near Nuneaton caused the temporary curtailment to Leicester from 12th December, however through services were resumed on 11th April 1949, but was soon cut back to Leicester. The motive power was usually an Immingham B1 and occasionally a K3 from Lincoln. From 25th September 1950 the Eastern Region altered the names of both Lincoln stations from MR to St Mark's and LNE Passenger to Central. As time progressed the former GC engines disappeared to be replaced by B1s, K3s and WDs. The first diesel multiple units 790XX series appeared on the Lincoln to Cleethorpes service in March 1955 and on

2nd July 1956 the service had become one of the first local services to become totally dieselised, the only exception was the Cleethorpes to Birmingham which remained steam until the early 1960s. The February 1956 '*Railway Observer*' reported that 9F 92035 was tested on the 4½ chain curve during December 1955, but proved unsatisfactory and were subsequently banned. The branch was served by a daily pickup goods up and down, in 1957 the northbound pickup goods left Holmes Yard at 10.40 a.m. and arrived at Barnetby 4.56 p.m.. The southbound goods left Barnetby at 9.40 a.m. and arrived at Holmes Yard at 4.37 p.m., crews usually were relieved at Market Rasen. These pickup goods also delivered cans of water to Level crossings No. 14, 18, 24, 31, 32, and 40.

The 1960s brought the end of local passenger and freight services. Claxby & Usselby station was the first to close to passengers and freight on 7th March 1960. Goods services were withdrawn from Howsham, Reepham and Wickenby on 30th December 1963. The following year North Kelsey lost its goods service from 25th May, and followed by Holton le Moor and Langworth on 15th June. Lincoln East Goods ceased to deal with public delivery from 3rd May 1965, however the private coal sidings continued. The Eastern Region proposed in late 1964 that passenger services between Lincoln and Cleethorpes should be withdrawn from 2nd November 1964, however objections were received and closure of local stations were not implemented until 1st November 1965, the last services being run on Saturday 30th October. The last timetable is given in Table 3.

The connecting curve from Newark Northgate onto the Midland line opened on 1st March 1965 and from that date the Belvoir and Highdyke Iron ore trains were diverted from the former GN Honington to Lincoln route, through Lincoln St Marks. The Belvoir traffic was diverted to Colwick and went forward on the 1550 7Y01 Colwick – Barnetby freight and Highdyke to Barnetby iron ore consisted of 4 trains per weekday.

Table 3

Weekdays			ASX	SO			B	
LINCOLN ST. MARKS			1016	1025			2016	
LINCOLN CENTRAL	0548	0831			1315	1717		2112
REEPHAM (Lincs)	0556	0839			1323	1725		2120
LANGWORTH		0842			1326	1728		
SNELLAND	0603	0847			1331	1733		2127
WICKENBY	0607	0851			1335	1737		2131
MARKET RASEN	0618	0902	1036	1045	1342	1744	2035	2137
HOLTON LE MOOR	0626	0911			1351	1754		2147
MOORTOWN	0630	0915			1355	1758		2151
NORTH KELSEY	0634	0918			1358	1801		2155
HOWSHAM	0638	0922				1805		2159
BARNETBY	0646	0930	1054	1103	1410	1813	2054	2207

A Derby Midland – Cleethorpes
B Derby Midland – Cleethorpes

Weekdays		A	B	SO	SX	C	D
BARNETBY	0714	0809	0909	1234	1234	1520	1806
HOWSHAM	0721			1241	1241	1527	1813
NORTH KELSEY	0725		0918	1245	1245	1531	1817
MOORTOWN	0728		0922	1249	1249	1534	1820
HOLTON LE MOOR	0732		0926	1253	1253	1538	1824
MARKET RASEN	0742	0830	0936	1304	1304	1550	1835
WICKENBY	0748		0942	1310	1310	1556	1841
SNELLAND	0752			1313	1313	1559	1845
LANGWORTH	0757			1318	1318	1604	1850
REEPHAM (Lincs)	0801			1322	1322	1608	1854
LINCOLN CENTRAL	0811			1333		1619	
LINCOLN ST.MARKS		0853	1000		1334		1905

A Cleethorpes – Nottingham Midland
B Grimsby Town – Derby Midland
C New Holland Pier – Lincoln Central
D Cleethorpes – Derby Midland

From 1st November 1965, the night shift between Spalding and Grimsby ceased, and the night shift commenced on the Lincoln to Barnetby line. Three Highdyke to Barnetby Iron ore trains were diverted and this virtually concentrated all Highdyke Iron ore trains onto this line.The increase in traffic determined that major track improvements had to happen. Each weekend from September to December 1966 work progressed on improving Lincoln to Market Rasen, the section to Barnetby to be dealt with in 1967.

Paytrain services were introduced from 29th June 1968 and Market Rasen became unstaffed for passengers, although the goods depot remained staffed. The closure of the former Great Northern lines in East Lincolnshire on 5th October 1970, caused all the Cleethorpes to London Kings Cross trains to diverted via Market Rasen and the Newark connection. From that day Market Rasen became the railhead for East Lincolnshire. The station was restaffed to deal with its new status, however the Goods depot closed from that day.

St Swithins Power station was extended in 1947 and 1957 by adding two 2 megawatt generators. In 1957 6 or

7 trips were booked from Holmes Yard to serve the station. It last generated electricity on 21st March 1977.

Inter City trains from Kings Cross to Cleethorpes, peaked with a service of 3 weekday trains and 2 on Sundays. The demise of loco hauled trains on the East Coast, affected this service and from 16th May 1982 it was reduced to one train eachway. Local trains now connected with the mainline at Newark Northgate, with 2 of the 3 diagrams were now loco hauled with a 31/4 and four Mark 1 coaches. High Speed Train (HST) commenced service on the sole Inter City train from 6th October 1982.

The new line connecting spur from the Midland railway onto the old Lincoln avoider at Boultham, onto the GN&GE Joint line at West Holmes, and then into Lincoln Central. This enabled the old line through St Marks to Durham Ox Junction to be closed. The new spur came into use on 12th May 1985, one day before the beginning of the new timetable. The last passenger to pass over the old line was the 1903 loco hauled ex Cleethorpes and 1818 HST ex Kings Cross. The empty stock from 2105 Derby–Lincoln St Marks was the last train to traverse the line from St Marks to Lincoln Diesel Depot. Durham Ox Junction to Lincoln Coal depot was subsequently singled.

The traffic over the curve from Durham Ox to Lincoln Central went from one train per day (Sundays excepted), 0545 Lincoln Central to Cleethorpes local (ex 0210 Manchester Piccadilly to Cleethorpes Paper/Mail train), to become heavily used. British Petroleum discovered oil between Lincoln and Market Rasen, and they built an oil loading terminal at Welton between Reepham and Langworth. The layout was brought into use during July 1985, and it is served by one daily train, which because of the track layout must run empty to Lincoln Holmes, before runround and backtracking to Langworth, and then reversing into the sidings. This train now forms the sole source of freight traffic generated on this line.

Locomotive hauled passenger trains between Cleethorpes and Newark Northgate, returned to be diesel multiple unit trains from 30th September, with the last trains hauled by 31441 1820 Newark to Cleethorpes and 31463 1930 Cleethorpes to Newark service.

Freight traffic largely consists of bulk oil and petrochemicals from Immingham and returning empty tanks. The Gainsborough to Immingham oil train is routed over this line. The connection between Durham Ox Junction and Lincoln Coal depot was taken out of use over the weekend of 20th January 1989, after seeing little recent traffic. Passenger services consist of one Inter-City 125 to and from London Kings Cross and a two hourly Provincial service from Birmingham to Cleethorpes formed of Sprinters.

The 1980's brought great change to the signalling, with signalboxes being closed, with Market Rasen on 22nd March 1981, Monks Abbey 23rd September 1984, North Kelsey 31st May 1987, Claxby March 1989, Moortown on 16th April and Snelland on 5th November 1989. All replaced by automatic barriers.

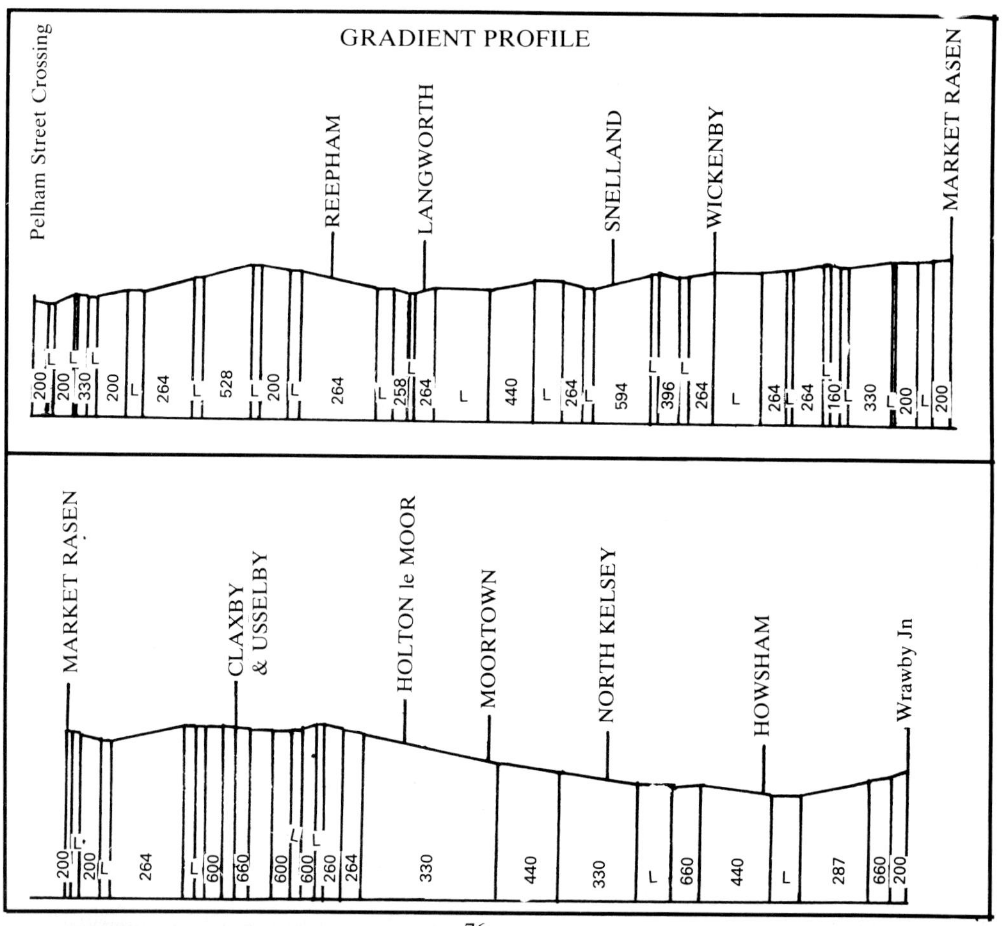

The 1305 Kings Cross to Cleethorpes train is eased round the corner from Lincoln St. Marks (portico visible behind the poplar tree) on 21st April 1981. During the last year of Deltic operation this train was a regular Deltic diagram. *(S. Turner)*

Although built for passenger work by the Great Eastern J69 68529 (converted to J69/1 in May 1954 and withdrawn in August 1958) is seen shunting wagons in the former Great Central Lincoln East Goods Yard on 7th April 1953. In the background the eastend of Central station is viewed with the end some coaches. *(J.R. Bonser)*

(below) Pelham Street level crossing at the turn of the century looking towards St Marks. It shows an unusual gantry taking the point and signal rods over the tracks and road. The signal on the right is not in use and would protect the GC connection to the High Street station of the GN. *(Jim Jackson Collection)*

Lincoln Central Platform 5 is host to a three coach 1.00 p.m. local to Cleethorpes on a misty 11th December 1954, headed by B1 61202. The platform awning covering the line between the two footbridges. *(T.J.Edgington)*

The 1607 Newark Northgate to Cleethorpes train hauled by 31434 stands in Lincoln Central on 26th September 1985. This was the last week of locomotive hauled passenger trains on this line. The stock is composed of four Mark 1's made up of 3 SK's and 1 BSK. *(R.E.Burdon)*

A three coach local passenger train leaves Lincoln, hauled by Beyer Peacock built J11, 5210 (became BR 64335 and was withdrawn in September 1954), with a Cleethorpes service in the late 1930's. *(E.R.Morten)*

On 8th September 1952, a J39/2 64906 heads a coal train from the Notts & Derbys coalfield through Lincoln Central onto the Market Rasen line. *(E.R.Morten)*

On 8th September 1952, a J39/2 64906 heads a coal train from the Notts & Derbys coalfield through Lincoln Central, onto Pelham Street level crossing, and takes the Rasen line. This crossing was replaced by a bridge and was offically opened on 27th June 1958 by H.M. The Queen. *(E.R.Morten)*

Class 114 diesel multiple unit headed by power car E56037, is about to pass Pelham Street Signal box with the 1302 Cleethorpes to Newark Northgate service on 21st August 1982. This was the only booked dmu diagram on the Cleethorpes to Newark Northgate service. In the background Lincoln Diesel maintenance depot is host to a few demic units an 03 mechanical shunter and an 08. *(S.Turner)*

Monks Abbey Signal Box (switched out) looking towards Lincoln on 29th January 1983. This used to be a busy box, with connects to the iron ore quarries to the west as well as Claytons and St Swithins Power Station. to the east of the line. *(Jim Jackson)*

L. N. E. R.
FOR CONDITIONS SEE BACK. Available for three days, including day of issue.
1598
REEPHAM (LINCS.) to
LANGWORTH
Fare THIRD / S \ CLASS
LANGWORTH

Reepham Station and Signal box in Victorian times, looking towards Langworth. Two station staff and the signalman are pictured as well as four children. The level crossing gates are made of iron and not wood as used today. Note the signalbox name being attached to the side of the cabin. *(H.Answer)*

A view from the down platform of Reepham, complete with a waiting shelter in August 1961. *(D.Thompson)*

On 29th September 1956, Langworth Signal Box looks newly painted. The crossing gates are of the single gate type (today it they are dropping barriers) The station consisted of the stationmasters house and booking office in the main building, a goods shed behind the house and a waiting shelter on the opposite platform. *(H.B.Priestley)*

A view of the rear of Stainton Crossing box on 7th April 1983. This was located between Langworth and Wickenby boxes and controlled the level crossing with protecting signals. This has since been replaced by automatic barriers. *(S.Gay)*

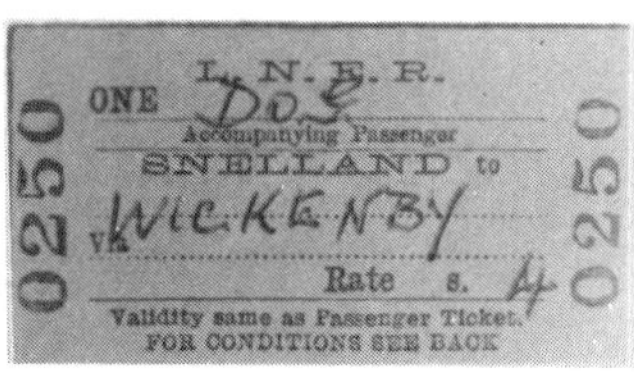

A lone man walks onto the down platform of Snelland station, light by oil lamps. The signal box can be seen on this side, while the stationmasters house and booking office are located on the opposite platform. *(D.Thompson)*

End on view of Wickenby signalbox, as seen on 7th April 1983. *(S.Gay)*

Wickenby station as seen from the level crossing, showing station masters house/booking office on the right. The platforms light by oil lamps. Note the station name is painted onto the top of the lamp. *(D.Thompson)*

B1 61408 (40B) heads a Football special from Grimsby to Lincoln through Wickenby on 29th Sept 1956. The locomotive supports a blue and white striped headboard, the team colours of 'The Mariners' *(H.B.Priestley)*

Market Rasen circa 1900, sees the arrival of a passenger service from Lincoln hauled by a Class 2 locomotive. The approach road is a hive of activity with nine passengers walking into town. Three cabs await custom outside the station. *(Lens of Sutton)*

The interior of Market Rasen station circa 1900. The original roof design of slate and a little glass made it fairly dark. A W.H.Smith bookstall was erected in the 1880's. Staff and a number of passengers await the arrival of a train from Lincoln. Note that the walls have enamel placks and advertise Suttons Seeds. Great Central posters publicise the companies shipping services to the continient. *(Heyday Publishing)*

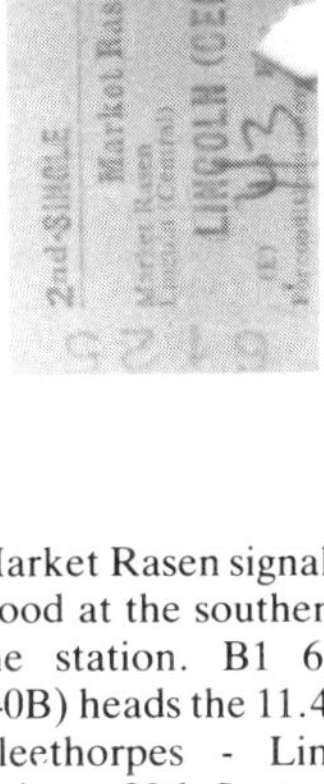

Market Rasen signal box stood at the southend of the station. B1 61406 (40B) heads the 11.46am Cleethorpes - Lincoln train on 29th September 1956. *(H.B.Priestley)*

The last B4 to survive 1482 'IMMINGHAM' stands at Market Rasen on 10th June 1947 with the 9.5 a.m. to Cleethorpes. Composed of three coaches and a parcels van. *(R.J. Buckley)*

31456 (IM) stands at Market Rasen with the 0545 Lincoln Central - Grimsby Town (ex 0210 from Manchester Piccadilly) on the last time this train ran as a passenger train. Note the small headboard 'The Last 1E62'. *(Geoff Hurst)*

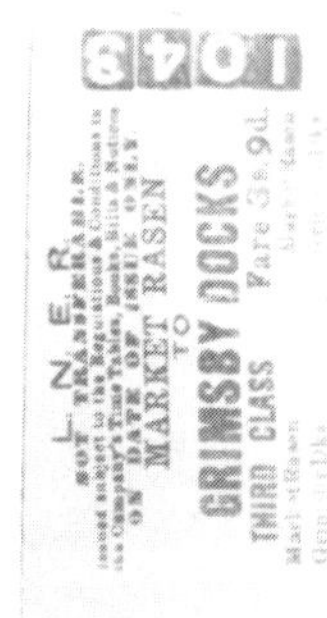

Only 15 months after closure the platforms of Claxby & Usselby have been cut back. The stationhouse now is a domestic residence. *(D. Thompson)*

WD 90510 (40B) approaches Claxby & Usselby with a long train of empties. 29th September 1956. *(H.B.Priestley)*

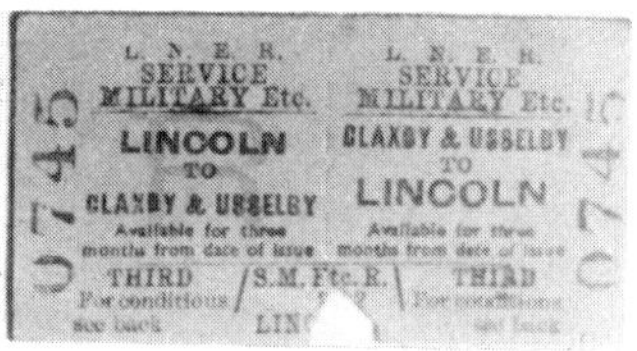

Holton le Moor signal box on 15th May 1982, winds the crossing gates by means of a wheel. The signals are now colour light and the gates are barriers. *(S.Gay)*

Holton le Moor station looking north in the 1930's with no waiting shelter on the down platform. The booking office/ stationmasters This station was different from all the others on this line, with the buildings being built well back from the level crossing. *(D.Thompson)*

J11 64305 (40B) leaves Holton le Moor with the down pick-up goods on 29th September 1956. Note the GN signal. *(H.B.Priestley)*

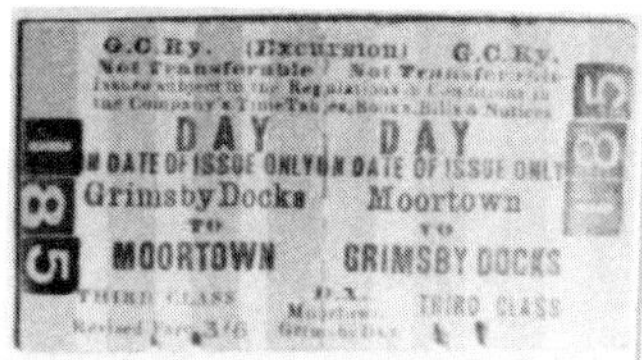

G.C.Ry. (Excursion) G.C.Ry.
DAY DAY
ON DATE OF ISSUE ONLY ON DATE OF ISSUE ONLY
Grimsby Docks TO MOORTOWN
Moortown TO GRIMSBY DOCKS
THIRD CLASS THIRD CLASS
3/6
0185

Moortown Signal box on 15th May 1982, operated a double gated level crossing. This has since been replaced by set of barriers. *(S.Gay)*

0127 2nd - SINGLE SINGLE - 2nd 0127
Moortown (for Caistor) to
NOTTINGHAM (MIDLAND)
(E) 10/6 Fare 10/6 (E)

Moortown station as viewed from the pedestrian crossing, during August 1961. A muck spreader sites in the goods yard on a flat wagon. *(D.Thompson)*

A southbound coal empties headed by O4/7 63758 (41J) sounds his whistle as its enters Moortown for Caistor station on 12th September 1959. Note the mobile crane and the fairly recent addition of a concrete shed. in the goods yard. Note the lamps are oil light. *(H.B. Priestley)*

North Kelsey Signal Box, Level Crossing and Station buildings as seen from the down platform, on 12th September 1959. *(H.B. Priestley)*

North Kelsey station in September 1961 as seen from the down platform. Note that up platform has no shelter. *(D. Thompson)*

WD 90696 coasts into Howsham with a down freight train on 12th September 1959. The station nameboard would be normally light by an oil lamp, but this one is derelict. *(H.B.Priestley)*

A local from Lincoln composed of a two car heavyweight and three car unit arrives at Howsham during June 1958. *(D.Thompson)*

(below left) Two Great Central Railway notices stand by a level crossing in Lincolnshire. *(Gordon Lacy)*

(below right) On 2nd May 1970, a fine gantry of signals (15-21) marks the end of the Market Rasen branch. The controlling box Wrawby Junction stands in the background. Today the gantry survives in a slightly reduced form, the Elsham and Brigg lines lost there gantries sometime ago. *(H.B.Priestley)*

5. Motive Power Depots

Inside Darnall repair shops stand O4/8 63882 and B1 61152 in 1954. *(Geoff Newall)*

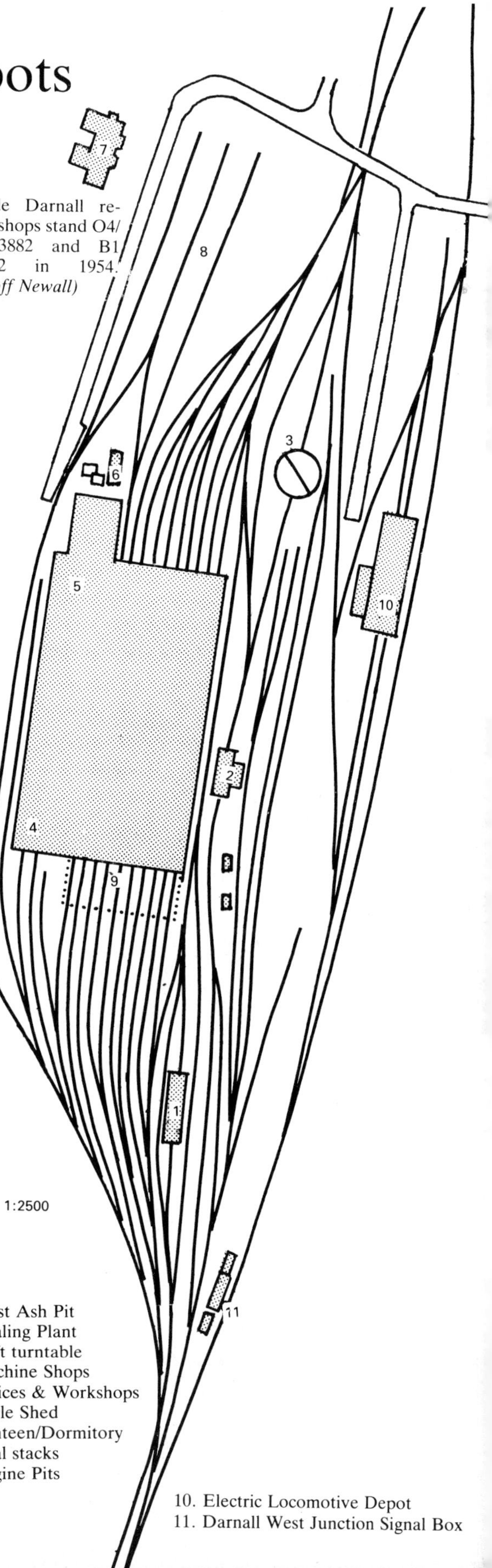

1. Darnall Steam Motive Power Depot

Traffic and Works Committee told on 1st May 1930, that Neepsend depot requires renewal of roof at an estimated £15,000. Neepsend shed limited to 5 acres expansion impossible, only 42 engines covered against proposed 112 at Darnall. Darnall shed to be built on down line near station with 70 foot turntable.

Darnall scheme abandoned 25th September 1930, due to low rate of Government assistance.

Contract No.2 for Darnall locomotive Depot, construct engine pits for steam engine shed, wet ash pit, foundations for engine turntable, wheel shop and stores estimated at £30,700, was awarded on 21st August 1938, to George Longdon for £23,390 subject to completion in 9 months.

Dormitory for enginemen, shed for locomotives, offices, works estimated at £71,660 awarded to George Longden £64,820 9s 2d 0 (error in original tender), complete in 12 months or amendment to tender.

Dormitories £5,679 19s 7d, Running foremans office £201 19s 6d, building for sand drier £276 18s 9d, cycle store £127 13s, urinal £52 12s 6d, shed for locomotives and offices £54,811 7s 4d; sundries (labour, bricks cement etc.) £3,556 14s 7d.

The commencement of hostilities in September 1939, caused all the contracts to be suspended with agreement with the contractors.

Construction of the steam depot commenced in 1942 and opened for traffic on April 10th 1943.

British Railways gave the depot the code 39B.

Depot code altered to 41A in October 1955.

Closed to steam from June 1963.

Retford Great Central shed, viewed from Thrumpton Level crossing footbridge in September 1963, shows the west end with two engines outside, a O4/3 63665 and an unidentified B1. *(P.H.Groom)*

Retford GC depot viewed from the London Road end on 4th May 1963. The turntable is immediately in front, with the brakedown crane and coach behind. The shed is well filled with locomotives. *(Geoff Newall)*

A fireman undertakes a very dirty job of shovelling out the ash, from the smokebox of a GN J6 64245 at Retford GC shed. *G.Coltas*

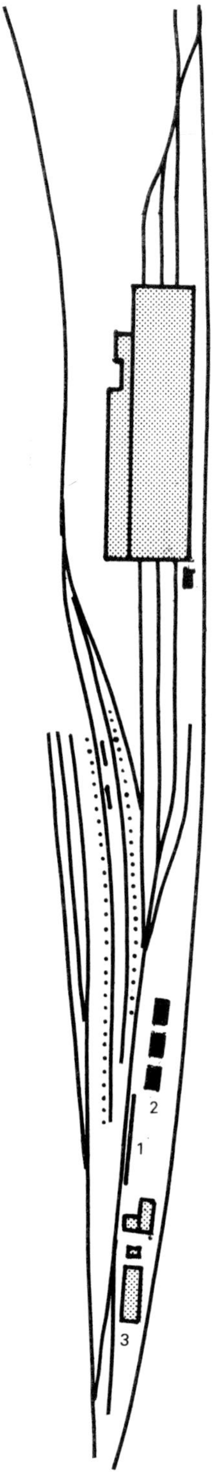

1. Fuelling Points
2. Fuel Tanks
3. Darnall West SB
4. Platforms ••••••

2. Darnall Electric/Diesel Maintenance Depot

Darnall Electric engine shed estimated at £69,349 plus C.M.E. £12,082 giving total of £81,831. This was revised on 21st July 1938 to £64,630, £13,725 and giving £78,355 respectively. The proposed length of Darnall Shed reduced from 404 feet to 300 feet.

A new contract was awarded by the Eastern Region on 13th March 1952 to Messrs G. Longden & Son Ltd for £12,936 6s 10d. Made up of 1½ tons of fabricated steel. 75,000 bricks, 290 cubic yards of aggregate, 120 cubic yards sand, 70 tons of cement. Contract No.8 Lines 17 feet 6 inches above rail level. Railway electrification Order 1932, states 20 feet, accepted 3rd March 1952. Time for completion to be agreed estimated £13,000. The contract for a small motor generator Messrs Mawdlings & Son Ltd £238 5s, was awarded on 21st March 1952.

Eastern Region announce plans to covert Electric Depot to Diesel in January 1958.

The 'Railway Gazette' gave the details. Adapted and enlarged for use. 300 feet long by 60 feet wide. Three roads and affording covered pit accommodation for 2 x 6 car dmus. Walls will be brick for first 4 feet, glass for next 22 feet. Power operated roller shutters, inspection pits between rails, side pits with lighting equipment. Workshops, offices, showers and staff mess room for 24 people. Possible to easily convert to electrics.

20th April 1959 Depot opened.

All mainline diesels transferred to Tinsley Diesel Maintainence Depot from 26th April 1964.

New depot code 41B, only diesel shunters allocated.

Lost all Diesel shunters to Tinsley from October 1965.

Diesel Unit maintainence transfeered to Tinsley in October 1984. Used by Carriage and Wagon department.

Carriage and Wagon Department closed at Darnall October 1987.

Used by track machines until October 1989 and is now semi-derelict

08878 stands with a coal wagon outside Darnall Diesel depot. All the roller doors are open and show the interior of the depot. The clock has been stuck at 1.57 for about 10 years. The depot was still in use to maintain diesel multiple units from Lincoln, Botanic Gardens, Neville Hill and Hammerton Street at night and weekends until October 1983. *(L.A.Nixon)*

3. Retford Motive Power Depot

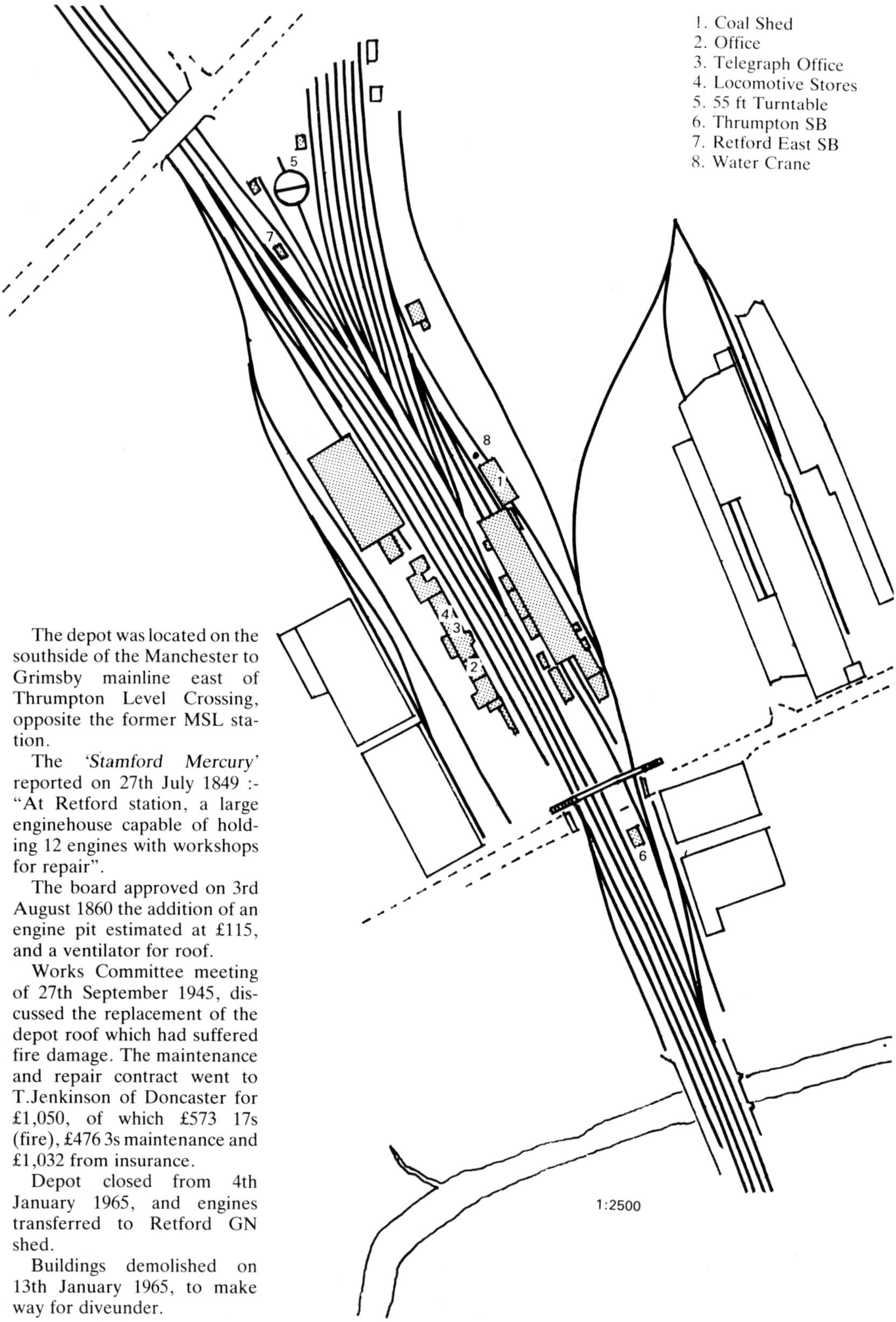

The depot was located on the southside of the Manchester to Grimsby mainline east of Thrumpton Level Crossing, opposite the former MSL station.

The *'Stamford Mercury'* reported on 27th July 1849 :- "At Retford station, a large enginehouse capable of holding 12 engines with workshops for repair".

The board approved on 3rd August 1860 the addition of an engine pit estimated at £115, and a ventilator for roof.

Works Committee meeting of 27th September 1945, discussed the replacement of the depot roof which had suffered fire damage. The maintenance and repair contract went to T.Jenkinson of Doncaster for £1,050, of which £573 17s (fire), £476 3s maintenance and £1,032 from insurance.

Depot closed from 4th January 1965, and engines transferred to Retford GN shed.

Buildings demolished on 13th January 1965, to make way for diveunder.

The eastern side of Darnall shed, as seen from the top of the coaling tower in 1954. Below nine locomotives composed of five classes :- an Austerity; B1's, O4's, a K3 and a J39. *(Geoff Newall)*

A view from the top of Darnall shed coaling plant in 1954, shows three classes of engine B1, O4 and K3, and the covered ashpit. *(Geoff Newall)*

Darnall on a Summer Saturday in August 1962, sees an English Electric Type 3 D6742 and unidentified Brush Type 2 arrive light from Sheffield Victoria, and descend towards the diesel depot. In the background a locomotive is on the turntable and only one engine stands outside the shed, however 8 of the 10 roads of the depot are occupied. *(Jack Davis)*

Great Central shed on 13th August 1936, with six engines, from left to right, 1946 numbers in brackets Ardsley K2 4673 (1763); March K2 4653 (1743); (front) Leeds D2 4376 withdrawn November 1937; (rear) B5 5185 (1688); B4 6096 (1481) and B5 6072 (1481) stands on the tank line with three wagons. Hidden inside are 12 engines made up of B4; B5; B7; D2's: D3; J6's and J11's. *(W.A. Camwell)*

Lincoln Diesel Maintenance depot on 7th May 1987, looking from the westend on a bright evening. Most of the allocation is hard at work, providing Lincolnshire and South Yorkshire with a very reliable service. These units travelled as far as apart as Norwich and Scarborough. Inside the depot is a 114 unit and a Cravens Driver Training Unit. The sidings hold two 114 units the first of which Lincoln depot had converted into a Parcels Unit by removing all but the end vestibule seats. *(Geoff Hurst)*

Inside Lincoln depot two units are undergoing maintenance, the nearest unit being a 114 (with a South Yorkshire PTE logo on the side) and a Cravens 105 converted into a driver training unit. *(Geoff Hurst)*

4. Lincoln (G.C.) Motive Power Depot

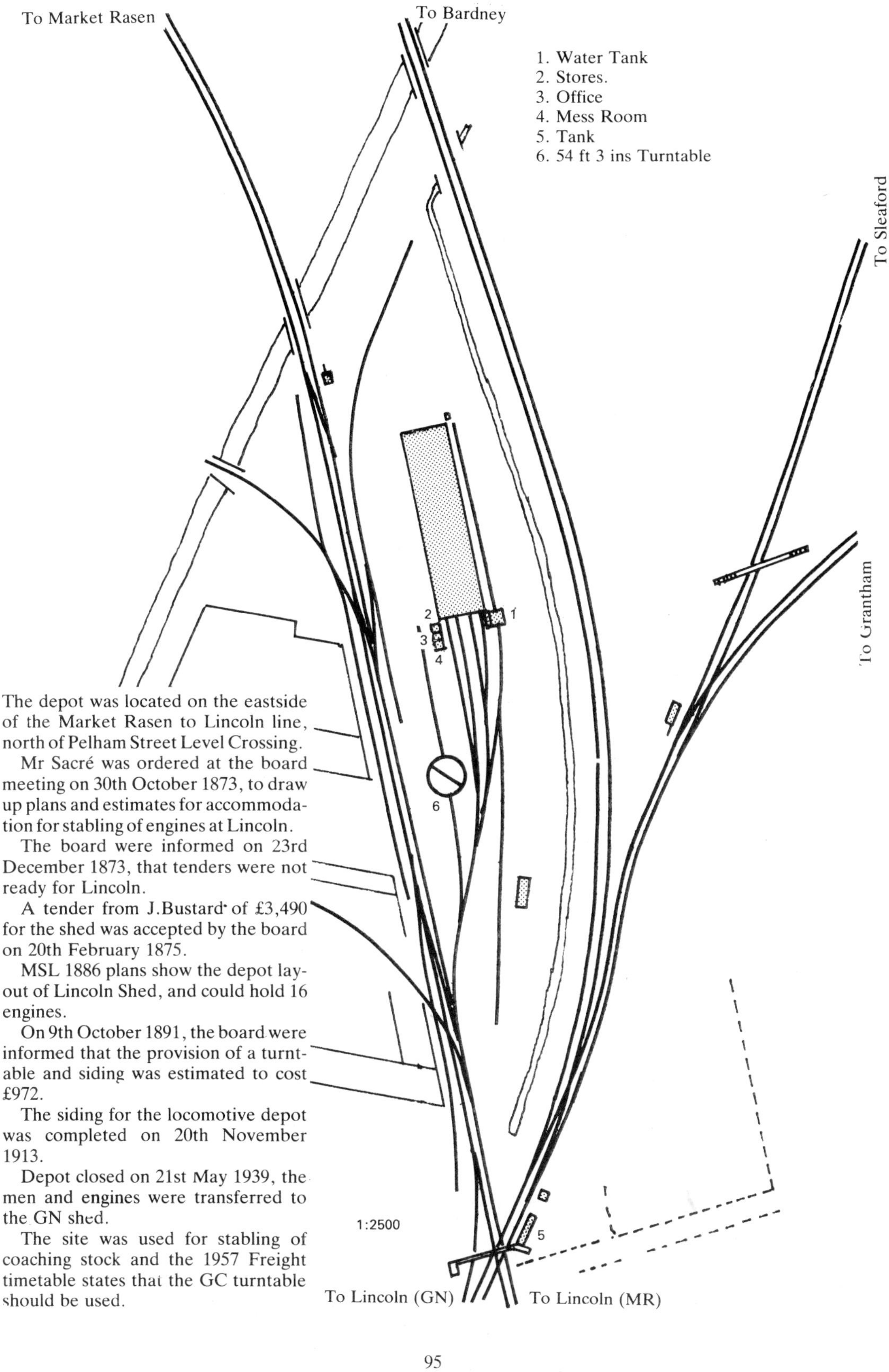

The depot was located on the eastside of the Market Rasen to Lincoln line, north of Pelham Street Level Crossing.

Mr Sacré was ordered at the board meeting on 30th October 1873, to draw up plans and estimates for accommodation for stabling of engines at Lincoln.

The board were informed on 23rd December 1873, that tenders were not ready for Lincoln.

A tender from J.Bustard· of £3,490 for the shed was accepted by the board on 20th February 1875.

MSL 1886 plans show the depot layout of Lincoln Shed, and could hold 16 engines.

On 9th October 1891, the board were informed that the provision of a turntable and siding was estimated to cost £972.

The siding for the locomotive depot was completed on 20th November 1913.

Depot closed on 21st May 1939, the men and engines were transferred to the GN shed.

The site was used for stabling of coaching stock and the 1957 Freight timetable states that the GC turntable should be used.

5. Lincoln Diesel Maintenance Depot

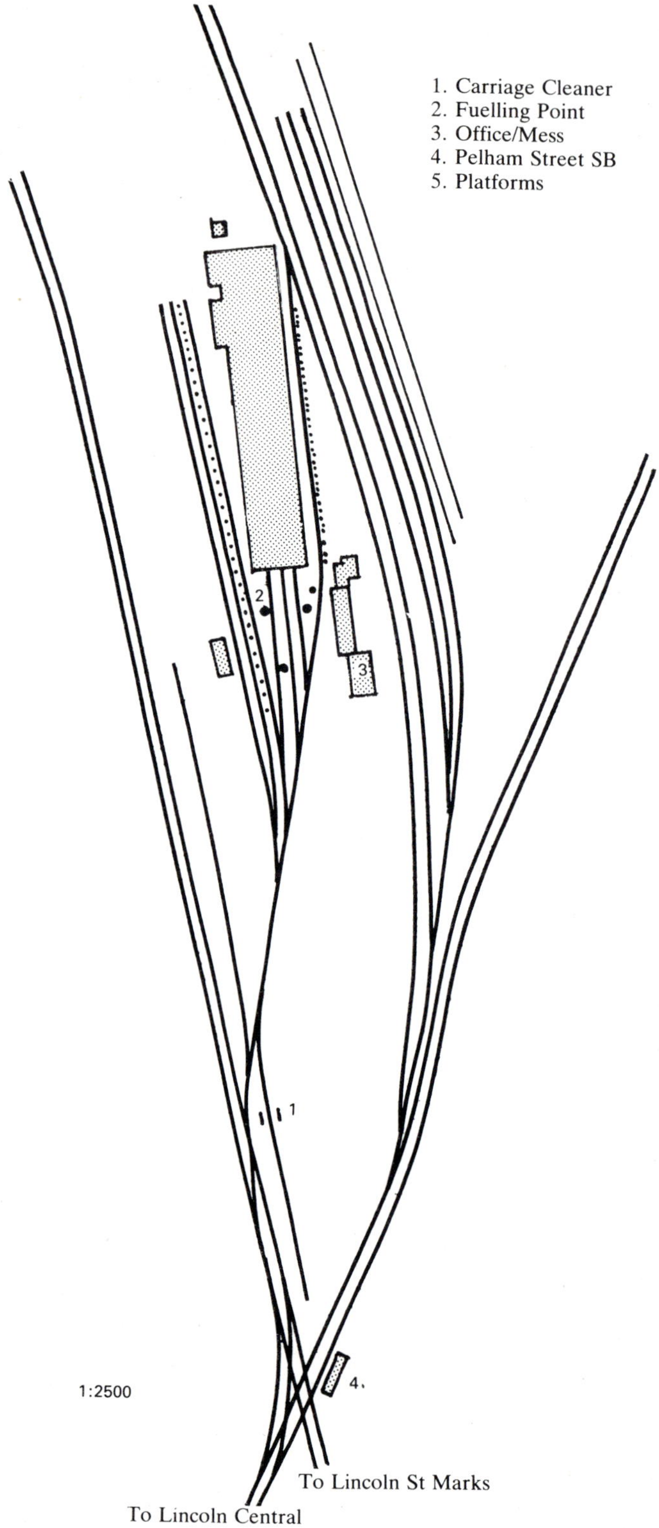

The diesel depot is was located on the eastside of the Market Rasen line north of Pelham Street Junction. Approximately on the site of the former MSLR shed.
In January 1955, the Eastern Region announced that it was going to build a diesel unit maintainence depot. required for new 2 car lightweight diesel multiple units for local services and also for diesel shunters. Estimated to cost £100,000. Steel frame shed covered with three roads, 2 of which equipped for pits for underfloor engines and transmission. One road will have additional 2 short platforms to enable work on shunters. 2 fuel oil tanks of 5,000 gallons.
Depot came into use late 1956/early 1957.
First mainline engines allocated in August 1963 :- D5801, D5802, D5803, D5853, D5855.
Remaining four Brush Type 2s 5689, 5801, 5802, 5803 transferred to Immingham 40B, May 1969.
The depot losses it allocation of shunters to Doncaster DR on 15th January 1986.
Last remaining Craven 105 unit withdrawn 11th May 1987.
The remaining Class 114 units transferred to Tyseley TS or Cambridge CA on 4th October 1987 and the depot closed.
R.F.S. Engineering Ltd of Doncaster lease depot in October 1989.